Quantum Metaphysics

To Felicity

Quantum Metaphysics

PETER FORREST

Basil Blackwell

First published 1988

Basil Blackwell Ltd
108 Cowley Road, Oxford OX4 1JF, UK

Basil Blackwell Inc.
432 Park Avenue South, Suite 1503
New York, NY 10016, USA

British Library Cataloguing in Publication Data

Forrest, Peter, *1948–*
Quantum metaphysics.
1. Metaphysics
I. Title
110

ISBN 0-631-16371-9

Library of Congress Cataloging in Publication Data

Forrest, Peter, 1948–
Quantum metaphysics/Peter Forrest.
p. cm.
Bibliography: p.
Includes index.
ISBN 0-631-16371-9
1. Quantum theory. 2. Physics – Philosophy.
3. Metaphysics. 4. Realism. I. Title.
QC174.12.F66 1988
530.1′2–dc19
88-10364
CIP

Phototypeset in 11/13 pt Baskerville
by Dobbie Typesetting Service, Plymouth, Devon
Printed in Great Britain by Bookcraft Ltd, Bath, Avon

Contents

Preface

Explanatory power, I believe, is a guide to truth. Quantum theory has immense explanatory power. But what is the truth to which we are guided? I am convinced that the proper method of answering this question is that of profuse speculation. We need speculations of all kinds, the fanciful and the down-to-earth, the radical and the conservative, the seemingly naïve and the seemingly oversophisticated. Then we should test them. Most will be rejected, not as falsified by experience so much as untenable. But a few will be retained. In this work I offer some speculations of this kind. I hope my example may encourage others to speculate further.

Most of my intellectual debts will be obvious, but I should mention three which are not so obvious. First, I would like to thank Henry Krips and Angus Hurst for their encouragement when I began thinking about these matters a few years ago. Secondly, I would like to thank David Armstrong. This might seem curious, since he has never worked in the area of quantum theory. But without his example as a metaphysician I would never have been able to engage in what is, as the title indicates, a work of metaphysics.

My thanks are also due to Cliff Hooker, Jack Lonergan and the publisher's referees for their most constructive criticisms of an earlier draft.

Hazel Gittins, Frances Redrup and Cheryl Chant deserve thanks for their work on preparing the typescript. Even if my writing were legible this would not have been an easy text to handle. Thanks are due to Elaine Leek, for her painstaking work as a sub-editor and to Steven Clarke for drafting the diagrams. Finally, I would like to thank Stephen Loveday for his help with proof-reading.

Introduction

I AIMS

This book is concerned with the metaphysical implications of quantum theory. I aim for the minimum of technicality, but even this minimum is more than I would like. To make it easier for the reader to omit technicalities at a first reading, I have relegated them to the notes. In addition, I have provided a glossary of technical terms.

The first three chapters are preliminary. In the first I say what I mean by *realism* about particles, or properties, or whatever. In the second I present an outline of such quantum theory as I need. In the third preliminary chapter I collect together various well-known problems and thought experiments, which provide the background for further inquiry.

In chapters 4 to 7 I consider various 'interpretations' about quantum theory, beginning in chapter 4 with the Copenhagen School, which I shall partially criticize and partially endorse. In chapters 5, 6 and 7 I discuss 'interpretations' involving particles of determinate location, waves and particles of indeterminate location, respectively. I conclude that there are at least three tenable positions, all of which are recognizably realist, as opposed to idealist or Kantian positions. I would have liked to have narrowed the range down to one, but this is not possible. In the final chapter I link up these speculations with other metaphysical issues concerning the status of persons.

My position is recognizably a realist one. As the proponent of realist speculations, I may face genuine difficulties. But there are also two quite spurious ones. The first is the natural, but scarcely rational, tendency to condemn a position not because of the *force* of the objections against it, but because of their *number*. The second spurious difficulty is the tendency to equate the tradition of Scientific Realism with Conservatism. I uphold the tradition of Scientific Realism not by returning to classical mechanics, but by *thoroughly* abandoning it. The insistence that realist speculations should be neo-classical is, I suspect, as responsible as anything else for the widespread abandonment of realist speculations about quantum theory.

II THE NEW METAPHYSICS

I am engaged on a work of metaphysical *speculation*. And I intend to be frank about it. Now the stereotype – a stereotype with a basis in fact – of the metaphysician is of a pompous and obscure dogmatist, who spins a bizarre theory out of principles which he – the stereotype is male – is convinced of, but no one else can even understand. Well, I reserve the right to be pompous. But what I am engaged on is what I call, with as much hope as faith, the *new metaphysics*. The new metaphysics uses argument without coercion. It is not dogmatic but frankly speculative. It has depth without obscurity.

First, consider argument. Nozick, whose approach to philosophy is on the whole congenial to me, rejects coercive philosophy. And he points out (1981, p. 4) that the metaphors we use are violent ones. He lists the *knockdown* argument, arguments which *force* you to a conclusion, ones that do not carry much *punch*, and so on. This theme could be elaborated. But elaboration is not necessary. What do these metaphors show? Do they show that philosophical argument is not 'nice', as Nozick curiously puts it (p. 13)? Nozick is on to something. But, please, let us not be 'nice'. Philosophers should be obnoxious – gently obnoxious. However, philosophers need not be coercive. In arguing we may and should seek to convince others regardless of whether they initially want to be convinced. We should realize, however, that the chances of winning such conviction are slim, and we should at all times show due respect. What is coercive is the use of derision (the raised eyebrow) or the appeal to fashion ('Since X's authoritative work no one can deny that . . .'). The genuine philosopher, while passionately arguing, even with raised voice and thumped table, never seeks to trick or deceive or bluff. Argument is neither rape nor seduction: it is honest proposal.

The new metaphysics is speculative rather than dogmatic. The proper method is to examine as wide a range of positions as possible, argue against some of them and defend others. Sometimes I judge that two positions cannot both be true.[1] But it does not follow that I have to decide between them. I would like to find a position which I could argue was the only tenable one. But very often I shall not find one. Quantum metaphysics exemplifies this. There are three positions which I consider tenable; two are particle 'interpretations', the other is a wave 'interpretation'.[2] I call this process of examining several positions, and arguing for and against, *speculation*. But it is similar to Nozick's *pluralism*. However, using that word would suggest that the multiplicity of tenable positions is a virtue. I would *like* to get a theory that far surpassed all rivals, although I see no need to deceive myself if I cannot achieve that goal. The term 'speculation' also suggests, as I want it to, a certain self-mockery. We should argue, and argue with passion; but all the while we should retain a sense of our finitude and historicity, knowing that what seems weighty to one thinker will not to another of the same generation, and even less to later generations.

Metaphysics is pointless if it does not aim for depth. There is a place for the detailed investigation of small topics. But metaphysics concerns big questions. I shall not here be concerned with the very biggest ones, such as 'Why is there something rather than nothing?' However, I am concerned with what quantum theory tells us about the world in very general terms, and so with questions such as 'What is the physical world composed of?'.

Quantum theory itself is somewhat speculative. It may or may not be replaced by some other theory.[3] So I am mounting speculation on speculation. My project, however, is not to assess quantum theory itself, but to rely on it to speculate about the physical world. And here I should say that I am not trying to give a metaphysical basis for any further progress there might be in physics. I aim to learn from modern physics rather than to teach.

I have been talking of the *new metaphysics* in the brash fashion of manifestos. But where are the new metaphysicians? I have already mentioned Nozick. Rather than list further examples I shall put the new metaphysics into perspective by sketching the history of metaphysics in the last forty years. Forty years ago metaphysics flourished among academic philosophers in India and among neo-scholastics, but had fallen into disrepute elsewhere. Anglo-Saxon philosophy was dominated by a mixture of Logical Positivism, Ordinary Language Philosophy and the philosophy of the later Wittgenstein. Therefore philosophers such as Anderson in Sydney who were interested in systematic metaphysics seemed to be living fossils. On the continent of Europe too, metaphysics was languishing. For the phenomenalist, hermeneutic and existentialist traditions have not encouraged metaphysical speculation, even though both Husserl and Heidegger had metaphysical interests.

Two movements have changed all this. First there is Scientific Realism, to be briefly discussed in chapter 1. Smart (1963) and Maxwell (1962) challenged the dominance of Positivism in the philosophy of science. Scientific Realism became a respectable *metaphysical* thesis. In the same tradition we have materialist theories of mind. Starting with the Identity Theory of Place (1956) and Feigl (1958), and developed by Smart (1959), Materialism was further elaborated by Lewis (1966) and Armstrong (1968), evolving into various sorts of Functionalism. Materialism, like Scientific Realism, is a metaphysical thesis.

The second great resurgence of interest in metaphysics resulted from the impact of semantics on metaphysics. Starting in 1961 with Quine's 'On What There Is', philosophers have become more aware of the ontological implications of their positions.

Both these influences have been of a non-dogmatic kind. Quine's account of ontological commitment freely admits rival paraphrases, and hence rival accounts of commitment. And the scientific realists and materialists typically relied on considerations such as Ockham's Razor (Smart, 1963, p. 11), which they admit to be inconclusive.

A further boost to the new metaphysics may turn out to be the attack, notably by Rorty (1979) on philosophy as the tribunal before which other disciplines must present themselves. This is significant because the role of tribunal has not been filled – at least not since Descartes – by metaphysics, but by epistemology or by philosophy of language. Metaphysics, although a branch of philosophy, was also expected to be judged by the tribunal, that is, by the dominant branch of philosophy. Hence the rejection of tribunals should liberate metaphysics.

III PURITAN ACCUSATIONS

I shall now imagine some intellectual puritan making five accusations against the new metaphysician. The first is that metaphysical speculation is fun. (It is also disciplined and therefore hard work, but the two are compatible.) No defence is required.

The second is that metaphysical speculation is frivolous: that is, it is useless, compared, say, to metallurgy. Even if it served no further purpose I would appeal to right-minded people to judge metaphysical speculation to be of *intrinsic* value. But is metaphysical speculation lacking in practical consequences? Traditionally the questions 'What is there?' and 'How am I to live?' have been linked. You should not seek Nirvana unless you accept some and reject other metaphysical theses about persons. You should not make your project in life that of loving God and your neighbour if you do not believe God exists. You should not, I say, give persons a privileged position in your ethics if you do not give them a privileged metaphysical status. And so on. Metaphysics has ramifications that touch on decisions. Moreover, within metaphysics one question leads to another. In particular, in chapter 8 I shall exhibit some of the ways in which quantum metaphysics affects issues to do with the status of persons. I claim, therefore, that some topics in metaphysics do affect decisions; and the rest of metaphysics is, in turn, relevant to these topics. I claim that practical decisions are made more wisely if they are made against a background of disciplined speculation of the sort I am proposing. In short: A leap of faith need not be a jump in the dark.

The third accusation is one of intellectual pride. 'Who am little I', the puritan asks – meaning 'Who are little you?' – 'to consider such matters?' I reply,'A human being.' However, I concede something. Pride is the besetting sin of philosophers, sometimes quite obvious, as with Herakleitos or Hegel, but at other times concealed by the appearance of modesty. And the typical expression of this pride has been the philosopher's inexhaustible powers of dismissal. Rorty draws our attention to this urge philosophers have to sit in judgment on other intellectual endeavours. I say philosophers should criticize other disciplines – and also take criticism from them. But that is the end of their judgment role. I propose as a prophylactic against the pride philosophers have too often shown that they be *frankly* engaged in speculation.

However, there is more to the charge of intellectual pride than simply pointing to a vice philosophers have often had. For a reading of Kant might suggest that all metaphysics is overweening pride. This is not the place for Kant scholarship, but his influence is strong enough to make some remarks on the Kantian threat appropriate. Fortunately, the threat is not based on the subtleties of his thought. I detect three lines of criticism of metaphysics stemming from him. The first is that metaphysical reasoning leads to antinomies. This suggestion is that for every argument put forward, say for the infinity of the past, there will be an equally good counter-argument. But an examination of Kant's antinomies will show that these arguments should never have been put forward as conclusive. Speculative metaphysicians find no scandal in equally good but opposing inconclusive arguments. Instead of lamenting the antinomies we seek further argument.

The second line of criticism is based on the form/content distinction, much used by Kant. And in spite of interesting criticisms by Davidson (1984, pp. 183–98), I uphold this distinction. We are then faced with the extreme difficulty of getting at content without form. This difficulty is interpreted as the unknowability of things in themselves. And, it is thought, it is the business of metaphysicians to know things in themselves. If so, metaphysics is white-anted.[4] The crucial step is the last. If I advocate, say, realism about particles, I am not claiming that talk of particles is the naked truth, that is the truth unconcealed by the forms provided by our thought and language.[5] Metaphysics is deep, but it does not have to be *that* deep. In this work I shall be concerned simply with what exists rather than with some category of the really real.

The third Kantian argument is based on the details of his account of knowledge. It is not just that we supply form to a given content. On Kant's theory, the form, not the content, is responsible for all those features that are knowable a priori. In particular the reliability of various rules of inference of a sort used in both science and metaphysics[6] would be ensured only by the fact that we give to experience a form which is in accordance with the rules themselves. It follows that the inferences would only be reliable when applied to the 'empirical' world which resulted from the giving of form to experience in this way. This would be a disaster, not merely for metaphysics but also for the more theoretical sciences and, in particular, for modern physics. But, in any case, Kant's argument is open to a well-known objection. Kant himself has described to us a process by which knowledge results. And this process is not just a matter of experience suitably formed. It is, to use Kant's magical term, *transcendental*. To rescue Kant from self-defeat it is necessary, I believe, to reconstruct Kant as *speculating*[7] about how we know, and as using that speculation to criticize claims to any metaphysical *knowledge* that is more secure than speculation. And *that* is no threat to the new metaphysics.

The fourth accusation is one inspired by Positivism. It is that metaphysics is meaningless. This charge lingers on even after verificationist theories of meaning have been abandoned as too stringent to allow even theoretical physics to have meaning. I have a suggestion as to why this should be. The charge lingers on, I submit, owing to the elasticity of the term 'meaning'. A relationship between people is said to be *meaningful* if it affects the people concerned. Again, if it is asked what the Budget *means* for people on an average wage, this is a way of asking how it affects them. The charge of meaninglessness is made, more often than not, against that which has no effect. Someone who says metaphysics is meaningless is probably not relying on the discredited verificationist theory of meaning at all. Rather she is criticizing metaphysics for being a piece of airy-fairy speculation removed from all practical matters. If so the charge is familiar enough. It amounts to the second charge again, that of frivolity. And this needs no further discussion.

The final accusation is that metaphysics is not to be understood literally but metaphorically.[8] I am prepared to grant that metaphysics is often metaphorical. But it is none the worse for that. My uses of the words 'particle' and 'wave' are certainly extended, and I have no quarrel with those who would call them metaphorical. However, if those are metaphors, then we are able to reason in metaphors. The belief that we are not, and hence that metaphors should be banished from philosophy, derives, I suspect from the assimilation of all metaphor to poetry; and metaphysics is not poetry, not even bad poetry.

IV INTERPRETATION AND SPECULATION

We talk of the Copenhagen Interpretation, or of a particle interpretation. What is it to *interpret* quantum theory? Is it like interpreting Aristotle, that is, making the obscure a little less obscure? A lot of that was done while quantum theory was being formulated, but I shall take it for granted. No. A particle *interpretation*, for instance, is the thesis that when a scientist, using quantum theory, mentions a particle, she is to be taken as saying that there is a particle and not just a 'particle'.[9] So it is a piece of metaphysical speculation not a piece of hermeneutics. It is the thesis that quantum theory describes particles. Likewise, a wave interpretation is the thesis that quantum theory describes waves. To call this *interpretation* is to perpetuate the confusion between the work of clarification of the theory, much of it ably performed by members of the Copenhagen School, and the work of making further metaphysical speculations about quantum theory. We would not talk of a materialist interpretation of the mental. Why then talk of a wave interpretation of quantum theory?

V IS INTERPRETATION NECESSARY?

Does quantum theory need 'interpreting', that is, speculating about? At first acquaintance it would seem so, for it is immensely puzzling. And it would make further acquaintance easier if we could interpret it in more or less classical terms, say as like classical statistical mechanics. And that, it could be urged, is what the felt need for interpretation amounts to – a desire for a neo-classical treatment. If so, it could quite properly be objected that those who persevere will become quite at home in quantum theory. So interpretation is unnecessary. Something like this argument is reflected in Heisenberg's statement that 'a real understanding [of the principles of the quantum theory] . . . is obtainable only through their relation to the body of experimental facts (1930, p. 66). This need not be taken as a statement of the positivist thesis that meaning is determined by verification procedures. We could take Heisenberg merely to be making the point that if you immerse yourself in the theory you will find it less puzzling.

This line of reasoning shows the need not to call speculation interpretation. For what Heisenberg has argued for, with some persuasiveness, is that quantum theory no longer needs interpretation in the sense of *clarification*. I say 'no longer' because we are considering the end product of a process of clarification. Indeed, those who, like Heisenberg, laboured hard on that process could justly feel irritated at the demand for interpretation. It is as if a very lucid writer were asked to explain himself further when all that is required is immersion in the text. But what I am after is metaphysics not hermeneutics. And immersion in the theory is not sufficient for metaphysical speculation about the further consequences of the theory.

NOTES

1 Typically if they have inconsistent consequences they cannot both be true. But there may be exceptions to this. See Priest (1979a) and Routley (1979).

2 In chapter 8 I discuss the issue of compatibility. I argue that variants of the two particle 'interpretations' are compatible.

3 Such as a hidden variables theory set up as a *rival* to quantum theory. This should be distinguished from hidden variables 'interpretations' of quantum theory.

4 Termites, often known as white-ants, eat wooden houses from within, leaving only wafer-thin shells.

5 Heidegger has been the champion of truth as the absolutely unconcealed ('alethia') (1959, pp. 102–3). If you like, I shall be operating with the concept of truth as the *relatively* unconcealed.

6 Roughly speaking, inference to the best explanation. See Forrest (1986) for some qualifications.

7 In the First *Critique* (B.xxvi, footnote) Kant opposes thinking to knowing. We can *think* things in themselves. This is not too far from an admission that we can *speculate* about them.

8 Current sensitivity to metaphors in philosophy owes much to Derrida (e.g., 1982, pp. 207–71).

9 If my use of 'particle' is already metaphorical, then the scare-quotes point to an even greater departure from the literal.

1

Realism

Realism is a compulsory topic in quantum metaphysics. We cannot go far without considering realism about particles or waves. Basically, I believe, realism is a simple matter, one of whether, say, particles exist. (Simple, I say, as if existence were simple, but let that pass.) Unfortunately philosophy is messy. I cannot bring myself to dismiss those who say: 'Particles exist. But they are not real, because their existence is just a matter of . . .'. Here the dots will be filled with some deflationary account of what it is for particles to exist.

In this chapter we shall get this messing about over and done with. The reader who is not concerned with recent philosophical discussions of realism may safely ignore the next three sections.

I ABSOLUTE AND RELATIVE REALISM

'I am a realist.' 'About what?' is the obvious reply. Realism is primarily relative. I concede the propriety of using it in a derived absolute sense to mean realism about THE WORLD, or something like that. However such absolute realism is not my concern in this book. So when I talk of *realism*, I mean realism about *X*s, for suitable *X*, as it might be realism about particles, or about properties. And, as a first approximation, a realist about *X*s is just someone who believes in *X*s, or, more precisely, believes there is at least one *X*. In that sense I am a realist about electrons, and about God, but not about phlogiston or the average plumber.

II BEING AND EXISTENCE

As a first approximation, a realist about *X*s is someone who believes there is at least one *X*. But Meinong and the neo-Meinongians[1] distinguish *being* from *existence* and tell us there are things which do not exist. And ordinary usage, for what it is worth, is on their side. Dragons do not exist. Mermaids do not exist. The fountain of youth does not exist. Square circles do not exist. So there are plenty of things which do not exist. I say, therefore, that a realist about *X*s is someone who believes at least one *X exists*. And I stipulate that both necessary beings and actual but contingent beings may be said to exist.[2] I further stipulate that non-actual beings do not exist.[3] I appeal to the reader to agree that these are felicitous stipulations in the context of quantum theory.

III DEFLATION

Recent philosophical discussion of realism has been dominated by Dummett's anti-realism and the rather similar Peircean Realism defended by Putnam (1978). First consider Dummett. In his 1963 paper 'Realism' (1978, pp. 145–65), he proposes two different criteria for realism. These are both of interest. So it does not matter much that he has subsequently clarified his position.[4] First we should notice that Dummett is concerned with realism in a more generic sense than is, or was, usual. He is concerned with what I shall call a *realist attitude* towards a class of statements (the disputed class), while I am concerned with realism about *X*s. I would say that realism is primarily an *existence* thesis, to use Devitt's helpful classification (Devitt, 1984). For Dummett, however, realism is much more of an *independence thesis* as Devitt puts it. Here is what Dummett says: 'Realism I characterize as the belief that statements of the disputed class possess an objective truth-value, independently of our means of knowing it: they are true or false in virtue of a reality existing independently of us' (1978, p. 146). Dummett's reference to a reality existing independently of us must be qualified. For we ourselves do not exist independently of ourselves and yet surely *that* does not prevent a realist attitude towards the statements asserting our existence. Accordingly I state the following independence thesis:

> A person is said to have an *independence-realist attitude* to a disputed class of statements if she *denies* that those which are true are true because of our means of knowing their truth.

Note the 'because' clause: I cash out independence in terms of explanation. This is in accordance with Dummett's use in various passages of the phrase 'in virtue of'. And it seems necessary if we are to define independence so as to cover even the case where, by good luck, we do know all the true members of the disputed class and there are no false ones which we think we know.

We also have in that passage a different criterion, stressed by Dummett elsewhere:

> A person is said to have a *bivalence-realist* attitude to a disputed class of sentences if she insists that every member of the class is either true or false.[5]

Only the first is an independence thesis, strictly speaking. As McGinn has argued (1982) and as Dummett has himself conceded (1982), these two sorts of 'anti-realism' in Dummett's papers are distinct.

Putnam's Peircean Realism is the claim that truth is that which would follow from an *ideal theory*. An ideal theory is one derived from the total evidence by the proper criteria for scientific theories. This concept of an *ideal* theory is not as clear as it should be. But let that pass. Peircean Realism could be restricted to a disputed class of statements, although that is not Putnam's intention. Furthermore, calling it Peircean *Realism*, although legitimate, could confuse. So I shall say that:

> A person has a *Peircean attitude* to a disputed class of statements if she claims that the members of the class which are true are true because they are implied by an ideal theory.

Peircean attitudes and the rejection of independence-realist attitudes are recognizably species of a genus of anti-realist attitudes. The generic attitude is that of assertibility-dependence, where:

> A person is said to have an *assertibility-dependence attitude* to a disputed class of statements if she insists that the true members of the class are true because of the way some specified actual or ideal human being or group of human beings could come to know them.[6]

Throughout the above I have considered *statements*, because I was using Dummett as a starting point. I do not think it will matter if we talk of sentences (more precisely sentence-tokens) instead.

What has all this to do with realism about *X*s? Suppose I am a realist about *X*s in the stipulated sense that I believe some *X* exists.

Someone of anti-realist leanings could concede that, but go on to say that *X*s only exist because of the way some specified actual or ideal group of human beings could come to know of their existence. Berkeley and his phenomenalist heirs provide an example. Phenomenalists do not deny the existence of material objects, but they insist that material objects only exist because we do or could perceive them. And perception, they would say, is our only way of knowing of the objects. Should we describe such a position as anti-realist? Contemporary philosophic usage suggests we should. However, *in this work* I prefer to call phenomenalists realists about material objects, but *deflationary realists*. I have three reasons for departing from the common philosophical usage. The first is that I prefer a threefold classification (into non-deflationary realists, deflationary realists and strong anti-realists) rather than a twofold realist/anti-realist classification. The second is that I hope some of my readers are physicists rather than philosophers, and I suspect that contemporary philosophical usage is confusing to those not immersed in contemporary philosophy. And thirdly, as the phrase 'Peircean Realism' shows, some measure of deflation is often considered compatible with realism, even in contemporary debates.

Deflationary realists about *X*s, then, accept that some *X*s exist but go on to say that they only exist because . . ., giving a *deflationary* account. By a deflationary account I mean one which, if correct, would result in a loss of significance or importance of the items being deflated. Typically, since the interest of metaphysics is primarily theoretical, the significance and importance of items is a matter of their explanatory role or, if this is more general, a matter of the role they play in our understanding of things. Deflationary accounts, then, eliminate the role these items play in our attempt to understand the world.

One deflationary account is that of an assertibility-dependence attitude. Even this, however, need not be deflationary. For if, like Dummett, and the recent Putnam, you have this attitude to *all* truths then perhaps it ceases to be a deflation of realism so much as a theory of truth. But at least an assertibility-dependence attitude to a *restricted* class is deflationary. Vague truths are an example. I say that vague truths are true because they may be asserted or accepted by those who know all the relevant *precise* truths. Since I do not have an assertibility-dependence attitude to all sentences, this is deflationary.

I have yet to discuss bivalence. And that is because it is not straightforwardly applicable to existence claims. If some *X*s exist, then the class consisting only of the single statement 'Some *X*s exist' contains a true member. The application of bivalence is not to the existence of some *X*s but to what we say about them. *X*s will be of a certain category, and as such there are various questions which we would normally ask about them. Thus, for particles we ask 'Where is it now?', 'What is its velocity?', 'What is its mass?'. We do not ask 'Is it sad?' or, literally, 'Does it know where it is?'. Let us form a class of statements consisting of answers to the questions which would be generally recognized as appropriate for *X*s. Failure of bivalence for this class is, I think, deflationary. Often this failure is a symptom of assertibility-dependence, for given assertibility-dependence there would often be no answer we could give to a question about *X*s even where it is appropriate. 'Is that man bald?' 'Yes and no' we reply, or alternatively, 'Neither yes nor no'. That is because, in some cases, our knowledge of all the relevant precise truths – we could have counted the hairs of his head – is not sufficient to decide.

It is widely thought that failure of bivalence is just a symptom of assertibility-dependence. But it can be of independent interest. Consider particles. Among the appropriate questions are 'What is its position?' and 'What is its momentum?'. Now if, as is often argued, quantum mechanical particles do not have both a determinate (i.e. sharp) position and a determinate momentum, then there is a failure of bivalence in the class of statements about particles. This could be taken as an indication of assertibility-dependence. But that is not inevitable. An alternative is to say that quantum mechanical particles are just not like our initial expectation of what particles are, so not all the questions that we thought appropriate are indeed appropriate. In that case we will tend to call them 'particles' rather than particles, or say they are not *really* particles. Either the scare-quotes or the 'not really' qualification is deflationary. We would be saying: Yes, particles exist, or at least something like particles, but they are not quite what we expect particles to be. In this way, failure of bivalence in the class of statements about *X*s can be a deflation of the realism about *X*s, without being a symptom of assertibility-dependence.

A further possible source of deflation is mentioned by Dummett (1978, pp. 156–60), that of *reduction*. He characterizes reduction as a species of, to use my terminology, assertibility-dependence deflation, and it is one he typically avoids. In a reduction of statements in the disputed class we require another class of statements, the reductive class. And someone who knew all the truths in the reductive class would then have the evidence needed to assert all and only the true members of the disputed class. I would add here that for a reduction the true members of the disputed class would have to be true *because* such a person would assert (or assent to) them. Otherwise, we have not excluded the case in which, by sheer luck, the evidence never leads us astray.

A reduction in Dummett's sense then, is a species of a deflation already considered, that of assertibility-dependence. But there is another sort of reduction. Sometimes the true members of the disputed class are true because of the truth of members of the reductive class, without the intermediary of a knower of the truths. Consider an atom. It exists because the electrons, protons and neutrons which make it up exist, have the properties they have and are related as they are. Can we not say that an atom is, for that reason, *reduced* to its constituents? I think we can, but there is nothing deflationary about saying an atom is just a system comprising electrons, protons and neutrons. This suggests that while reduction may be accompanied by deflation, there is always some other source of the deflation, typically assertibility-dependence.

It might be objected that a reduction of persons to systems of molecules is deflationary. Here I reply that if it is deflationary, and materialists may deny it is, it is not because of the reduction but because of a strong anti-realism accompanying the reduction. We tend to think of persons as being more than their bodies. Let us call this something extra, whatever it might be, the *spiritual*. Then even a materialist who identifies the mental with the physical, and so is a realist about the mental, is a strong anti-realist about the spiritual. And strong anti-realism is as deflationary as you can get.

I shall continue, then, to say someone is a realist about *X*s if she believes some *X*s exist. But I concede that in addition to the strong anti-realist who denies this, there will be deflationary realists, such as Dummettians, who accept realism but with a deflationary gloss. One sort of deflation is the assertibility-dependence of the existence of *X*s. Another is failure of bivalence in the class of statements about *X*s. And there may be others, although I do not include reduction as an independent kind of deflation.

IV SCIENTIFIC REALISM

My speculations about quantum theory are against a background of Scientific Realism. But what is Scientific Realism? When Smart (1963) and Maxwell (1962) argued for it in the early sixties, they were arguing for the existence of electrons and other subatomic particles (or 'particles'). They were not arguing for the existence of all the theoretical entities postulated by all sciences. Nor were they even arguing that we should take the whole of their preferred scientific theories literally. Smart, for example, explicitly excepted lines of force (1963, p. 34). As I understand it, their aim was to argue for the existence of certain selected theoretical entities, and in so doing, overthrow Positivism and Instrumentalism, which were then the orthodoxy.

This poses a problem. Can Scientific Realism be characterized more precisely? Boyd's formulation, as quoted by Putnam (1975, p. 73) includes the claim that 'terms in mature scientific theories typically refer'. Will that do, at least as a necessary condition for Scientific Realism? I think not. First, mature scientific theories include terms that ostensibly refer to mathematical entities. And it would seem a further metaphysical position, not itself part of Scientific Realism, to believe in those. More serious is the reliance on the phrase 'mature science'. What does a body of theorizing have to do to become mature, and what entitles it to the honorific 'science'? One suspects that mature sciences are precisely those that the scientific realist accepts. I believe in God. Why should I not take theology to be a science? And if it is, then it is indeed mature.

Since I am discussing quantum theory, I can avoid these problems of characterizing mature science. For my purpose, what is of interest is realism about quantum theory. Is this, then, the claim that the terms of that theory typically refer? Surely not. For one of the terms of that theory is 'particle'. But whether or not the term 'particle' refers or, better, applies to anything is a matter for speculation even among those who consider themselves scientific realists.

Boyd's formulation is not especially helpful. Let us turn, then, to van Fraassen.[7] He formulates Scientific Realism, which he opposes, as the position that 'Science aims to give us, in its theories, a literally true story of what the world is like; and acceptance of a scientific theory includes the belief that it is true' (1980, p. 8).

Van Fraassen has shown us one way of not being a scientific realist, namely accepting a theory but not believing it to be true. There are others. A religious fundamentalist might reject (and so not *accept*) every piece of science she has heard of, and yet believe that the aim of the sciences was to achieve literal truth. Such a person is recognizably even less a scientific realist than van Fraassen.

Restricting our attention to quantum theory, should we say, then, that someone has a scientific realist attitude to it if she believes it to be true? That will not do, for under a phenomenalist interpretation, quantum theory is treated as a systematization of our sensations. So a phenomenalist could believe it to be true, without believing in any theoretical entities. Van Fraassen's qualification 'literally' serves to exclude the phenomenalist. But it also excludes a wave interpretation. For a particle is not literally a wave.

We could go on trying to characterize a scientific realist attitude to quantum theory, as the holding of some thesis. But I think we should try a different approach. Scientific Realism was not, I submit, a thesis at all. Rather it was a *movement* characterized in three ways:

(1) By a lack of deference to Positivism and Instrumentalism which were then dominant.
(2) By realism (in my stipulated sense) about some, selected, theoretical entities, especially electrons, protons and neutrons. (It is rash to attempt to state a general thesis here about the entities the scientific realist is a realist about.)
(3) By reliance on two, rather different, argument-schemas. One, which I reject, is the continuity argument based on the lack of a precise boundary between the observable and the non-observable.[8] The other, which I accept (with qualifications that need not concern us here[9]) is inference to the best explanation, as Harman calls it. (Ockham's Razor plays a supporting role here.)

In my study of quantum theory I bring two things from the Scientific Realism movement. The first is the lack of deference to Positivism and Instrumentalism. This lack of deference is now the orthodoxy among philosophers, but less so among physicists. The second is a respect for inference to the best explanation. Both these are implicit in the new metaphysics. Clearly the non-dogmatic speculative approach I have been advocating precludes all deference to any philosophical orthodoxy. So even if Positivism and Instrumentalism were still dominant, that would not concern me. And inference to the best explanation, or some argument-schema like it, is essential to the new metaphysics if it is to be anything more than merely a list of possible theories. The new metaphysics is a matter of disciplined speculation, not a mere exchange of opinion. And without some such argument-schema there would be no discipline, beyond perhaps[10] the sorting of speculations into the consistent and the inconsistent. Furthermore, although many actual speculations are inconsistent I have yet to meet one that could not be made consistent with a few dodges.[11] Hence something like inference to the best explanation is essential for the new metaphysics.

This is a good place to discuss an objection to my whole project. To state it I begin by describing a well-known children's game. I say '*Simon says* stand on your head and wiggle your toes'. If you do not, then you are *out*. But if I now say 'Stand on your head and wiggle your toes', without the prefix 'Simon says', and you do, then you are *out*. Here is a variant called *Science Says*. If the accepted science says *p*, then you believe *p*; if not, then you don't. Scientific realists do not have to play Science Says. Science does not tell the whole truth, and, anyway, they can have reservations about many scientific theories – much of science is speculation. The objection that I need to consider is that quantum theory is too speculative to be accepted, and so, by relying on it, I am playing Science Says.

The obvious reply to the objection goes like this: Of course quantum theory may be wrong, but it is interesting to discuss certain matters within the scope of the supposition that it is right. The obvious reply does not satisfy me. Speculation is not discussion for its own sake. It is speculation about the truth. What I say, rather, is that the product of my speculations will be quantum theory plus further metaphysical considerations. The argument for any such speculation is based on the explanatory power of the total speculation. But much of this explanatory power is the same regardless of which further metaphysical considerations we add on to quantum theory itself. The complete discussion of my proposals would also include the class of rivals to quantum theory, such as hidden variables theories. Hence my project is incomplete; it is part of a wider project without which it would lose its point. But such incompleteness is inevitable. For one consideration leads to another and a *complete* piece of metaphysical speculation would cover just about every topic anyone has ever discussed. Nor is this incompleteness an especially severe one. For what drives people to consider rival theories is the belief that quantum theory however 'interpreted', that is, with whatever metaphysical supplement, is untenable. So by showing that there is at least one tenable 'interpretation' I shall remove one motive for hidden variables theories. It remains, of course, an open possibility that some approach incompatible with quantum theory will provide better predictions than quantum theory and so be preferred to it. In that case all my work will have been in vain. But we always take *that* sort of risk.

Respect for the Scientific Realism movement does not, then, require us to play Science Says. Indeed it is not clear at the outset quite what it does require of us in the case of quantum theory. Conceivably we could end up with no tenable speculation. In that case, we could do no more than note statistical correlations between data (Ballentine, 1970). At the other extreme we might hope for a metaphysical theory which was as satisfactory as the natural or common-sense understanding of Classical Mechanics. My speculations fall between these two extremes. I hope I provide more understanding than is given merely by a combination of mathematical formalism and statistical data. But none of the speculations I have provided is quite as satisfying as we might have hoped.

V ON PROPERTIES

Many, otherwise well-educated, people are largely ignorant about medieval thought. But one thing they do know is that there was a dispute between realists and nominalists, which concerned, in part, realism about properties thought of as universals, that is as capable of belonging to more than one item. I sympathize with non-deflationary realists about universals, being a student of Armstrong (1978) in that respect. It might be suspected then that I am covertly appealing to such realism. This suspicion will occur especially when, in chapter 7, I provide a theory of indeterminate properties. Am I not assuming realism about properties? No, but I assume something. And this is a convenient place to discuss *what* I assume. It is that there is a prima facie case for properties, a case which can be met, if at all, only by systematic ways of paraphrasing those statements which are prima facie about properties. Consider the Pap/Jackson argument elaborated by Armstrong (1978, vol. 1, ch. 6). Truths such as 'Orange resembles red more than it resembles blue' seem to be about the properties *orange*, *red* and *blue*. Nominalists need some strategy for paraphrasing such sentences to remove the prima facie commitment to properties. I say to them: Paraphrase away, but paraphrase too my references to properties. In this way I avoid any controversial realism about properties.

The nominalist will attempt to paraphrase *all* statements which are prima facie about properties. But even the realist may want to paraphrase *some* of them. For a start, in this book I treat location (or position – I do not distinguish the two) as a property. It might be claimed that it is more naturally treated in some other way.[12]

We might also worry about *momentum*. Here it is convenient to make a distinction between the *occurrent* and the *dispositional* momentum. I shall describe them for the case of a particle in Classical Mechanics. (It is an open question whether, according to quantum theory, a particle *has* momentum, in one or both senses.) A classical particle has a centre of mass $q(t)$ at each time t. If it is a point particle then $q(t)$ is simply its location at time t. Provided it moves smoothly we can define its occurrent momentum as its mass times the rate of change of the location of its centre of mass ($m \times \mathrm{d}q/\mathrm{d}t$). Thus, its occurrent momentum at time t is defined in terms of its locations at times near t. Even if its motion is not smooth, we may still be able to define an approximate occurrent momentum which is the mass times the average rate of change of location for a suitably small time interval.

We might well be suspicious about the occurrent momentum. If location is not a genuine property, could something defined in terms of it be? By contrast, I have no scruples about treating the dispositional momentum as a genuine property, if we need it at all. This is the disposition or tendency a particle has to move. The connection between the two senses of momentum is that a particle with a dispositional momentum p and subject only to finite forces will, as a result, move in such a way that the occurrent momentum equals p.

VI REALISM ABOUT QUANTUM SYSTEMS

Guided by the Scientific Realism *movement*, we might note the immense explanatory power of quantum theory and declare ourselves realists about its theoretical entities. Now there will be considerable dispute as to whether, say, particles are theoretical entities of quantum theory. But there will be much less dispute about whether *systems* are theoretical entities of quantum theory. For the theory explicitly concerns quantum systems, whether they turn out to be collections of particles, waves or something else. And quantum systems have *states*.

Therefore I declare myself a realist about quantum systems and their states. And in accordance with my realism about properties, I take *being in a given state* as a property of the system – or, on some speculations, a property of a collection of systems. (But, as I have said, nominalists may paraphrase.)

This realism about quantum systems and their states is defeasible. If weighty objections can be found to it, then we should abandon it, in favour perhaps of van Fraassen's acceptance without belief. My chief thesis in this book is that there are no weighty objections to realism about quantum systems and their states. Indeed in chapters 5, 6 and 7 I give three different tenable accounts of what quantum systems are.

Here the realist may wonder why we should not be *primitivists*, that is take the systems as unanalysable entities. To be sure we talk of *one-particle* systems, *two-particle* systems, and so on. But there are ways of characterizing the number of 'particles' in a system without invoking a further analysis of the system as actually composed of particles.[13] The most obvious way of developing this primitivism about systems is to adopt the semi-Kantian 'interpretation' according to which the system is something characterizable only by its propensities to affect us in various ways. I discuss, and reject, this in chapter 4. I require, therefore, that there be *some* further account of what a quantum system is. It does not follow, however, that I am proposing a hidden variables *rival* to quantum theory. I am not. The further account is a metaphysical description of the quantum system, not a rival scientific theory.

VII STATES AND SUB-STATES

The states of a quantum system are mutually exclusive.[14] That is, I take it, part of what we mean when we talk of *states*. But in any case it follows from the usual formalism for quantum theory. So we may talk of *the* state of a quantum system. However, there is a possible confusion here, for we shall have occasion to consider *sub-states* of a quantum system. For the sake of definiteness, consider the following example. Chapter 5 concerns an 'interpretation' in which a quantum state describes, among other things, a statistical distribution for the location of particles in a large collection. Suppose the particles in the collection have names, 'particle-one', 'particle-two' etc.[15] Then the very same statistical distribution would result if we swapped particle-one and particle-two. So the one quantum state would be the disjunction of very many sub-states. Each sub-state could be described by saying where particle-one is, particle-two is, and so on.

Confusion could occur if we talk of *the* state of the system. Do we now mean the quantum state, which is a disjunction of sub-states, or the sub-state? To avoid this confusion, I stipulate that by *the state* we mean the quantum state. The sub-states of quantum states are to be called sub-states.

NOTES

1 Notably Routley (1979) and Parsons (1980).

2 Contrast Heidegger and the existentialists. They restrict existence (Heidegger's *Dasein*) to a contingent mode of existence, exemplified by human beings.

3 Contrast David Lewis (1966, p. 3), who insists that non-actual possible things exist.

4 In a statement of his position in 1982 Dummett notes how in the past he has emphasized bivalence, but that now he considers the issue to be rather subtler. None the less he still thinks bivalence is important.

5 Dummett presumably means the disjunction to be inclusive, so that believing a statement to be both true and false is not a rejection of bivalence. For he considers bivalence to be one of four semantic principles paralleling four logical principles (1978 xix). And the parallel for bivalence is $A \vee \neg A$, where 'v' stands for inclusive disjunction. However, believing a statement to be both true and false will often be quite as anti-realist as believing it to be neither. The position that there are true contradictions (i.e. some truths are false) is called di-alethic logic. It has been championed by Priest (1979a) and Routley (1979). Dummett, understandably, would not seriously have considered it.

6 I restrict this to *human* beings, because the thesis that truth is what is asserted by God does not seem to be an anti-realist one.

7 Van Fraassen's formulation is similar to Hooker's (1979).

8 There is no sharp boundary between the observable and the non-observable. But there is no sharp boundary between belief and disbelief either. The two 'sway together' as David Lewis would have put it. Hence there is no continuity problem in believing in the observable and disbelieving in the unobservable. (See van Fraassen, 1980, p. 214).

9 For my criticisms of inference to the best explanation see *The Dynamics of Belief* (Forrest, 1986, ch. 10). However these criticisms are minor ones. I grant that the features which give a theory explanatory power also provide support for the theory.

10 Since di-alethic logic (belief in true contradictions) is a tenable speculation, I do not think we could even eliminate inconsistent theories.

11 Such as distinguishing respects. If something is both Φ and not-Φ then all we have to say is that it is Φ in one respect but not-Φ in another.

12 One line of thought here (Armstrong, 1978, vol. I, pp. 118–25) is that the position of an object is its *particularity* or *thisness*. If so, it is not a property in the usual sense.

13 We require more than merely the system and its states in order to characterize the number of 'particles'. What we need is an action of the group of symmetries of space–time on the 'space' of all states of the system. If the latter is represented by unit vectors in a Hilbert space, then any connected group of symmetries will be represented by unitary operators on the Hilbert space. Roughly speaking, a one-particle system is given by an irreducible representation. More generally we can consider the algebra of those projection operators which commute with every unitary operator representing a symmetry of space–time. The number of generators of this algebra is the number of 'particles'. (With an irreducible representation the algebra is generated by the identity operator.)

For a discussion of the characterization of a single-particle system see Piron (1976, ch. 5), and the works he cites.

14 If they are time-dependent states, they are mutually exclusive *at a given time*.

15 Some may doubt whether it is possible to *name* the particles. I remind them that this is just an illustration of a sub-state. It need not be taken too seriously.

2

An Outline of Quantum Theory

The theory whose metaphysical implications I shall investigate is the orthodox quantum theory of a finite number of particles in the non-relativistic limit. This orthodoxy is itself partly the work of the Copenhagen School,[1] but may be distinguished from the metaphysical speculations – including anti-metaphysical speculations – of that school. That both formulation and metaphysics get called *interpretation* is unfortunate. For, by accepting the Copenhagen formulation I might be accused of prejudice. I plead not guilty. The Copenhagen formulation is simply what most physicists now mean by quantum theory and those who reject it have the task of formulating rival theories.

In this chapter I present the amount of quantum theory which I think necessary for quantum metaphysics. Because I shall present it with the minimum of technicality, the discussion here is not a complete introduction to quantum theory.[2]

I A SINGLE SPINLESS PARTICLE

Quantum theory is complicated by the mystery of *spin*, and by the subtlety of systems of many particles. So, as is customary, I begin with the simplest case, that of a single spinless particle.

I divide the theory of a single spinless particle into three parts. First I discuss the range of possible states of a one-particle system, then the dynamic constraints on the system, and finally the measurement theory for the system.

First, then, consider the possible states of the quantum system. I shall here be considering the *pure* states as opposed to the *mixed* ones, which I discuss later. The possible states may be said to form a state-space. This 'space' is not literally a space or space–time. It is a space only in the metaphorical sense of being a suitably structured set of states.

Following von Neumann (1955), it is now customary to represent this structure using an infinite-dimensional Hilbert space. Every pure state then corresponds to a one-dimensional subspace of this Hilbert space.[3] For our purposes, however, we may rely on the finite-dimensional and, in particular, the three-dimensional analogue.[4] We may think of the states as like the rays of ordinary three-dimensional space. Take some point *O* as fixed. Then consider all the doubly infinite straight lines going through *O*. By a *doubly infinite* line I mean one that goes on indefinitely in both directions. These I call the *rays* through *O* (see figure 2.1). Notice that since a ray is doubly infinite, it has no preferred direction, so moving a ray round by 180° results in the same ray.

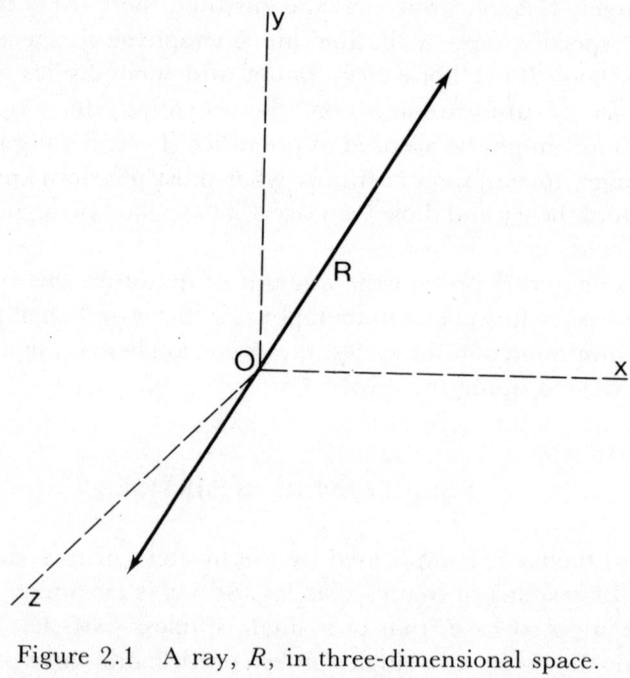

Figure 2.1 A ray, *R*, in three-dimensional space.

The structure of the space of states could be derived from the 'angles' between states,[5] which are visualized as the angles between the rays. The maximum angle between two rays, namely 90°, is especially significant. States making this 'angle' are said to be *orthogonal*.

Another important relation is that in which a third state *s* lies in the 'plane' to which two other states *u* and *v* belong.[6] In that case, *s* is said to be a *superposition* of *u* and *v*. This is illustrated in

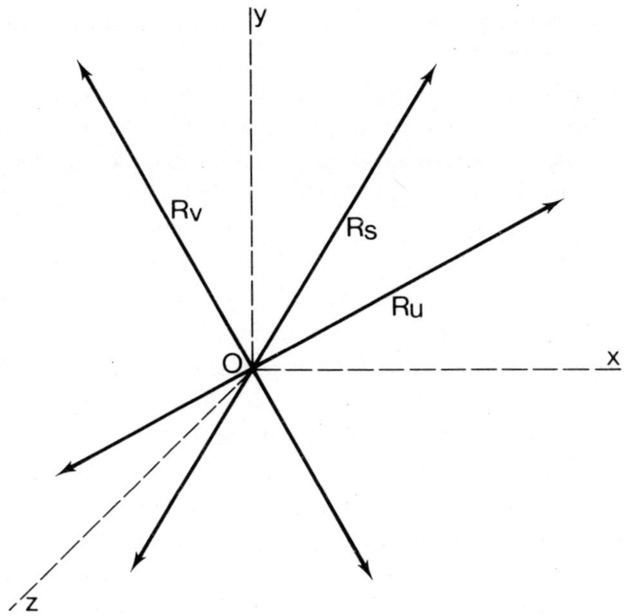

Figure 2.2 A ray, R_s, representing a superposition of two states u and v.

figure 2.2, where the ray R_s is in the plane to which R_u and R_v belong. We may think of the superpositions of u and v as being *between* u and v.

Given two states u and v we may ask whether there are any non-trivial superpositions of u and v. (The trivial ones are just u and v themselves). If not, then we have what is called a *superselection* rule. But such rules are exceptional. Typically, for every ray in the plane to which R_u and R_v belong, there will be a state which is a superposition of u and v.[7]

That states typically have superpositions is the source of many of the differences between classical and quantum mechanics. Consider, for instance, the twin-slit thought experiment (discussed in detail in chapter 3). An electron is forced to pass through an apparatus containing two slits. If we had observed the electron going through one slit, or if we had shut the other, then the state would be u. Likewise if we had observed the electron going through the other slit then the state would be v. In classical mechanics the state of the system would have to be either u or v even if we had no information about which slit the electron went through. But in quantum theory the state will typically be a superposition of u and v, which is a third state identical to neither u nor v.

How does the state change over time? Here I distinguish what I call the Schroedinger dynamics from what I call the von Neumann dynamics. The former governs the behaviour of idealized systems not subject to observation or any other external influence. The latter concerns the mysterious 'collapse of the wave packet' on observation. I use the phrase 'von Neumann dynamics' rather than the more familiar 'collapse of the wave packet'. For the latter is too closely tied to a particular interpretation of von Neumann dynamics.

I shall discuss the von Neumann dynamics in a while. But I first consider the Schroedinger dynamics, in which quantum states evolve deterministically.[8] That is, so long as there is neither outside interference nor observation, the state at time $t + \Delta t$ is completely determined by the state at time t. In this respect the state of a one-particle system in quantum theory is analogous to the pair consisting of both position and momentum for a particle in classical mechanics. The precise way in which the state evolves depends on the forces operating on the particle. This is part of what physicists labour to calculate. But it does not affect our metaphysical discussion.

In our three-dimensional analogue we can think of the ray R moving but with O fixed (see figure 2.1). The Schroedinger dynamics preserves the 'angle' between states u and v. That is, if u and v make a specified 'angle' at one time t, they evolve into states making the same 'angle' at a later time. In particular, orthogonal states evolve into orthogonal states. Yet again, if s is a superposition of u and v then s^*, the result of evolution of the state s is a superposition u^* and v^*, the results of the evolution of u and v.[9]

If only there were no von Neumann dynamics. How simple quantum theory would be! But any theory describing the world in unfamiliar terms, whether it be that of Parmenides or that of Heisenberg, should link the unfamiliar to the familiar. And this we have not yet done. To be told that there are these states which evolve in a deterministic fashion unless interfered with, is curiously frustrating if you are interested in, say, an electron. We want to know where the electron is, or how fast it is moving. Quantum measurement theory provides the requisite link with the familiar by telling us what is likely to result if we make various observations or measurements. First it should be noticed that quantum measurement theory concerns only highly idealized measurements. And these measurements are *doubly* idealized. Not only do we have the inevitable idealization that various sources of error may be ignored, we also consider measurements which involve only a single *instantaneous* interaction with the system by the apparatus performing the measurement. Otherwise, the Schroedinger dynamics of the system under the influence of the apparatus would affect the result. Actual measurements, then, will only approximate those of measurement theory. So it is no criticism of quantum theory if there is no apparatus which could exactly produce quantum measurements. A further point to be noted is that we should not assume at the outset that the measurement is *veridical*, that is that even these idealized measurements or observations give us truths about the particle. That is a matter for further discussion.

The basic observables for a single spinless particle are position and momentum (or, since we assume the mass is known, velocity). Ordinarily we think of an idealized measurement of position as telling us a precise location in space, the *observed position*. Likewise, we think of the idealized measurement of momentum as resulting in a precise value, the *observed momentum*. But, for a variety of reasons, it is convenient to concentrate for the moment on idealized Yes/No observations (YNOs).[10] Let me emphasize that not every Yes/No observation is a YNO, only those idealized ones for which the von Neumann dynamics holds. If we measure position, say, then we shall be able to answer various Yes/No questions such as 'Is the result of the measurement in region V?'[11] To answer that question is to make a position YNO. Or again, if W is a set of momenta we can find out whether the result of measuring the momentum is in W. This is a momentum YNO.

Each YNO A has a negation $\bar{A}$. $\bar{A}$ gives the result No just when A gives Yes, and vice versa.

The von Neumann dynamics is specified by the *eigenstates* (elliptical for 1-eigenstates) of the YNO. The set of eigenstates Eig(A) consists of all those states for which there will be a Yes result with probability one. So Eig($\bar{A}$) consists of all those states for which there will be a No result with probability one, and hence a Yes result with probability zero.

If observation were as in classical mechanics, then every state would belong either to Eig(A) or to Eig($\bar{A}$). But things are more complicated. We find that:

(i) Every member of Eig($\bar{A}$) is orthogonal to every member of Eig(A).

and:

(ii) Every state is a superposition of some member of Eig(A) and some member of Eig($\bar{A}$).

If we had a superselection rule telling us that there were no non-trivial superpositions of members of Eig(A) and Eig($\bar{A}$), then this would be the classical situation. But typically we do have non-trivial superpositions. And for these states there is a probability strictly between 0 and 1 of a Yes result. The von Neumann dynamics is specified by stating: (a) what this probability is; and (b) how the state evolves if there is a Yes result, and how it evolves if there is a No result.

Both (a) and (b) may easily be described in terms of the *projection* on to Eig(A). This projection is defined as follows. Given a state s which is neither in Eig(A) nor in Eig($\bar{A}$), there is a unique pair of states u and v such that u is in Eig(A), v is in Eig($\bar{A}$), and s is a superposition of u and v.[12] Then u is the projection on to Eig(A) of s, and v is the projection on to Eig($\bar{A}$) of s. Another way of defining the projection is to say that the projection of s on to Eig(A) is that

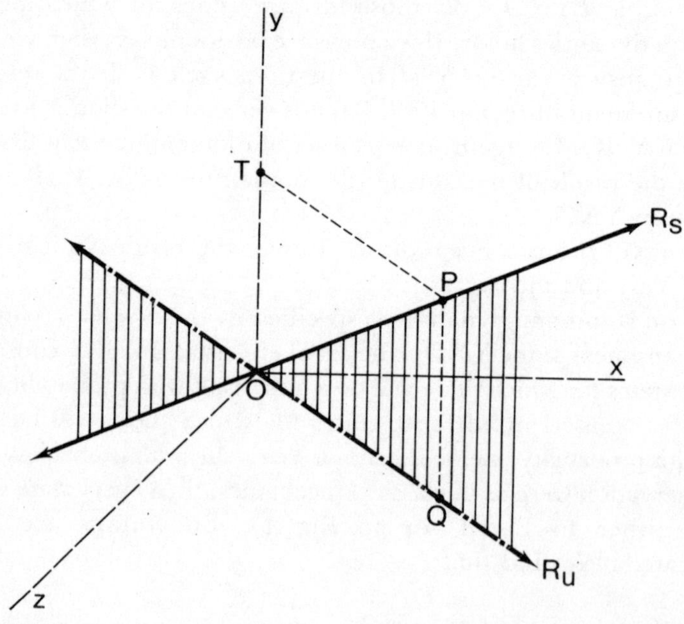

Figure 2.3 The projection of a ray, R^s, on to the Oxy plane.

member of Eig(A) which makes the smallest 'angle' with s.[13] We may visualize the projection operator by considering the three-dimensional analogue. Suppose the eigenspace Eig(A) consisted of all the rays in the Oxz plane (see figure 2.3). We can project down on to the Oxz plane by 'shining a light from above or below'. The 'shadow' of a ray R_s down on to the Oxz plane will be a ray R_u, the projection of s on to Eig(A).

If the result of the YNO A is Yes, then the state changes to its projection on to the eigenspace Eig(A). Likewise, if the result is No, the state changes to its projection on to Eig($\bar{A}$). And the probability of a Yes result is (cosine γ)^2 where γ is the 'angle' between the state and its projection on to Eig(A). This can be interpreted without using trigonometric formulae if we consider a vector OP of unit length along the ray R_s. If we project down (see figure 2.3 again), then the 'shadow' of OP is a vector OQ where Q is perpendicularly below P. Then the probability of a Yes result is the square of the length of OQ, which we write as $|OQ|^2$.

Likewise, the projection on to Oy, which in this example is the unique member of Eig($\bar{A}$), is the vector OT. So the probability of a No result is $|OT|^2$. By Pythagoras' Theorem, $|OQ|^2 + |OT|^2 = |OP|^2 = 1$, so the probabilities add up to one, as they should.

The von Neumann dynamics, then, is specified quite straightforwardly by the projections on to the eigenspaces. It should be noted that a YNO is *repeatable* in the sense that if it is performed again, without any lapse of time, then we get the same result. So if the previous result was *Yes*, the new one will (with probability one) be *Yes* again. If the previous result had been *No*, the new result would (with probability one) be *No* again.

This follows from the formalism because if the state is already in Eig(A), as it is after a *Yes* result, it remains unaltered when the YNO A is performed again.

Von Neumann dynamics is probabilistic unless the state is an eigenstate either for A, the YNO in question, or $\bar{A}$ its negation. For in other cases there is a probability μ strictly between 0 and 1, of transition to some specified eigenstate for A, and a probability $1 - \mu$ of transition to a specified eigenstate for $\bar{A}$.

What sort of probabilities are these? That is a matter for speculation. But this much seems clear: there must be some connection between probabilities and frequencies. Suppose we had a large number of systems, each of the same sort and each in the same state. And suppose in each case we measured the same YNO, say whether or not the particle is in V. Then there will be some frequency f, of cases in which we get a Yes result. And the theory gives us a probability μ for this outcome. We should not expect that $f = \mu$. For a start, as Reichenbach has pointed out, f will be a fraction whereas μ may well be an irrational number. The Law of Large Numbers only tells us that, in suitable circumstances, it is *improbable* that f differs much from μ. That is of no help if we do not yet know what sort of probability, and hence improbability, we are considering. However, we do know that, except as a result of the acceptance of quantum theory, the only guide to these probabilities are the corresponding frequencies. So unless we are fairly confident on any given occasion that the frequency will approximate the probability we have no grounds for accepting quantum theory at all. Hence the probabilities mentioned in quantum theory, whatever they turn out to be, must be related to frequencies in the following way:

> For any given observation and any given state of a given system, if the observation were repeated often, we should predict that the relative frequencies of results would approximate the probability of the result.

It might be objected that by sheer bad luck the frequency might not approximate the probability. To this I say: Yes, by bad luck what we should predict may turn out to be false. But this does not show we should never have made the prediction. For example, consider a lottery with a thousand tickets. I should predict that ticket number 231 will not be selected. Yet some ticket will be selected. So one of a thousand predictions of this sort will be incorrect.

Although the YNOs are treated as basic in the formalism, we may also consider observables which take a continuous range of values. In particular, we may consider the observation of position and the observation of momentum. Suppose we observe that the location is q to within a small error δ. Then that is equivalent to getting a *Yes* result to the YNO of whether the particle is in a sphere centre q and of radius δ. The probabilities for (approximate) location are conveniently described by a *probability density function* ρ, which is the probability per unit volume. The sphere of radius δ has volume $\varepsilon = (4/3)\pi\delta^3$. So the probability of the location being q to within δ is approximately $\varepsilon \times \rho(q)$. The smaller δ is, the better the approximation.

What we cannot do, given only the standard formalism,[14] is to handle *exact* observation, such as that the particle is exactly at q. But this need not worry us. For exact observation would be a further idealization, and one we are not required to make.

Likewise there is a probability-density $\hat{\rho}$ whose values are such and such probability per momentum cubed. The probability of obtaining the result that the momentum is p to within an error of η units of momentum is $\zeta \times \hat{\rho}(p)$ where $\zeta = (4/3)\pi\eta^3$. Here ζ is just the 'volume' of the sphere of 'radius' η centre p.

I now turn to Heisenberg's justly celebrated uncertainty principle. The probability densities for position and for momentum enable us to calculate the *mean* position and the *mean* momentum. The significance of the mean position is that in a large number of repetitions the average of the position obtained will approximate this mean. We then measure departures from the mean for position by the square of the difference between the result of the measurement and the mean. This quantity itself has a mean, which is called the *variance* for position. Similarly we can define the variance for momentum.

The variance is a measure of how widely we should predict the results of a large collection of observations to be scattered. The smaller the variance the less they scatter. Heisenberg's Uncertainty Principle now tells us that in the standard units (for which Planck's constant is 2π) the product of the variances for position and momentum is at least 1/4. Thus the less the predicted scatter for position the greater it must be for momentum.

II THE WAVE REPRESENTATION

In section I, I presented a fairly 'abstract' version of quantum theory. However, the working physicist likes the more 'concrete' wave-mechanical approach in which states are represented by suitable *functions*. If we stick to a single spinless particle the functions concerned have as values complex numbers. That the functions are complex-valued need not unduly concern us. For the most part we may think of them as *like* functions whose values are just the familiar real numbers.

The probability-density is given by the simple formula:

$$\rho(q) = |\Phi(q)|^2$$

Here ρ is the probability density and Φ the wave function. $|\Phi(q)|$ denotes the absolute value of $\Phi(q)$. If Φ is a real-valued function, then $|\Phi(q)|$ is just $\Phi(q)$ made positive, that is it is $\Phi(q)$ itself if $\Phi(q)$ is greater than zero and $-\Phi(q)$ otherwise. In that case, then, $|\Phi(q)|^2 = \Phi(q)^2$.

Every wave function $\Phi(q)$ has a Fourier transform $\hat{\Phi}(p)$ which is a function not of q, the position, but of p the momentum. The non-technical idea behind the Fourier transform is that we can analyse the wave into sinusoidal (i.e. wavy) waves of varying wave lengths each corresponding to a given momentum.[15] The Fourier transform $\hat{\Phi}(p)$ tells us the result of this analysis. $\hat{\Phi}$ is calculated from Φ using a straightforward mathematical formula.[16]

The probability-density for the momentum is given by:

$$\hat{\rho}(p) = |\hat{\Phi}(p)|^2$$

This is similar to the formula connecting $\rho(q)$ and $\Phi(q)$. Indeed, as far as the mathematical formalism goes, we could represent the state as easily by $\hat{\Phi}(p)$ as by $\Phi(q)$. However, since position seems more fundamental than momentum, I shall consider the state to be represented by a function of position.

The eigenstates for the YNO of whether the particle is in region V are just those wave-functions Φ for which $\Phi(q) = 0$ if q is outside V. Likewise, the eigenstates for the YNO of whether the particle has momentum in W are just those wave functions Φ, for which $\hat{\Phi}(p) = 0$ if p is outside W.

The 'angle' between u and v, and hence the orthogonality of u and v, is easily expressed in terms of the wave functions representing the states u and v.[17]

Superposition is especially simple in the wave representation. For if Φ_u represents u and Φ_v represents v, then any superposition of u and v will be represented by a constant times Φ_u plus a constant times Φ_v. Thus we simply add up suitable multiples of the two functions to get the superposition.[18]

The Schroedinger dynamics is given by the Schroedinger equation, a differential equation for $\Phi_t(q)$ (representing the state at t), which is deterministic. That is, for functions $\Phi_t(q)$ satisfying the equation, $\Phi_{t+\Delta t}(q)$ is determined by $\Phi_t(q)$.

The importance of the wave representation is that it suggests one of the 'interpretations' to be discussed, according to which a quantum system is a field, and the state of a system is indeed a wave in the general sense of being a disturbance to the field. This approach is a matter of taking the complex numbers as representations of the possible values of a field at a given location. Then the wave function represents the values of the field at each location.

III SPIN

Most particles have intrinsic angular momentum, or spin, as it is usually called. This is a vector quantity which is added to the extrinsic angular momentum (namely that due to the particle's velocity) so as to give the total angular momentum. If the particle were actually spinning, the angular momentum due to the spin would be such an intrinsic angular momentum, and its direction would specify the axis of rotation. But the amount of the intrinsic angular momentum for an electron is larger than one would expect from the literal spinning of a particle.[19] So when I talk of *spin* I simply mean intrinsic angular momentum, without specifying any underlying explanation.

Spin has two (other) peculiarities. The first is that its magnitude is constant, for a given kind of particle. Only its direction varies. I shall concentrate on particles such as electrons, protons and neutrons which have spin one-half (in the usual units in which Planck's constant $= 2\pi$). So the spin is a vector of length one-half pointing in any direction.

The other peculiarity is that, unlike position or momentum, we cannot measure the spin with arbitrarily great accuracy. All we can do is make various YNOs which I shall now describe. These YNOs are all of the same form, they are the observations whether the particle's spin is 'up'. If it is, then the answer is Yes. If not, the answer is No. What counts as 'up' is relative to a direction, namely the direction which is to count as 'straight up'. So for each direction the associated YNO has a Yes or an 'up' answer just in case the direction of spin makes an acute angle with that direction, and a No or 'down' answer just in case the direction of spin makes an obtuse angle. We may ignore the infinitesimal probability that the direction of spin is at right-angles to the specified direction (see figure 2.4).

These YNOs are analogous to the YNO of whether a particle is in region V. But while we can have the latter YNOs for any size of region, those for spin all involve a region of spin-directions comprising half of the possible directions. And since the measurement of spin

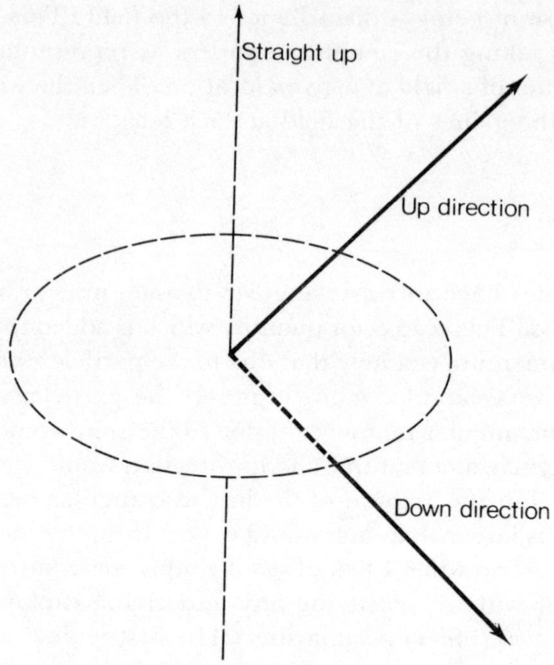

Figure 2.4 Up and down spin directions.

typically affects the spin, we cannot obtain greater accuracy by repeated measurements.

The probability that the result of making a spin YNO is 'up' (for a given direction of 'straight up') is specified in the usual way by the eigenstates for that YNO, that is the states which give an 'up' answer with probability one.

There is a neat formula connecting the angle between the straight-up direction for two spin YNOs with the 'angle' between the eigenstates. Suppose the straight-up directions make an angle α. Then given any eigenstate s for one of the spin YNOs, there is a smallest 'angle' between s and any eigenstate for the other spin YNO, and this 'angle' is $\alpha/2$. So if s is an eigenstate for the first of the two spin YNOs, and we then make the second YNO, the probability of a Yes result is (cosine $\alpha/2)^2$ (Messiah, 1962, p. 546). Now every state is a superposition of eigenstates for spin YNOs. So we obtain the further consequence that the only states orthogonal to an eigenstate for a given spin YNO, A, are the eigenstates for $\overline{A}$. In other words, if s is a state which gives an 'up' result with probability one, the only states orthogonal to s are those giving a 'down' result with probability one with respect to the same 'straight up' direction.

An analogue of Heisenberg's Uncertainty Principle is worth noting. If we make a spin YNO (or any other YNO) there are only two results, Yes and No, obtained with probability λ and $1 - \lambda$ respectively, for some λ. We assign the number 1 to Yes and 0 to No. Then we can calculate the mean. It is just λ, the probability for Yes. And the variance is $\lambda(1 - \lambda)$. It can be shown that if we have two spin YNOs whose straight-up directions are at right-angles, then the sum of the variances is at least 1/4.[20] So not both variances can be less than 1/8. This limits the possibility of reducing the scatter in repetitions of experiments that measure spin.

In the wave representation of a single particle with spin one-half, the function will no longer be a complex-valued function but a *spinor-valued* function. Spinors are represented by pairs of complex numbers. Hence a spinor-valued function may be thought of as a pair of wave functions of the sort considered in the previous section. But there is another way of looking at it. To each spinor z we can associate a unit vector σ in three-dimensional space.[21] If the spinor-valued function is $\Phi(q)$ let the associated vector be $\sigma(q)$. Then the direction of $\sigma(q)$ can be thought of as a direction of spin at the location q. The eigenstates for a given spin YNO are represented by the spinor-valued functions for which $\sigma(q)$ has always the same direction (where it is non zero). This direction is then the straight-up direction for the spin YNO in question.

If we think of the representation of the state as a disturbance to a field then the direction of $\sigma(q)$ at q is analogous to the direction of polarization of an electro-magnetic wave. And it is having this direction of polarization which distinguishes the wave representing a state of a particle with spin from one representing the state of a spinless particle.

IV TWO-PARTICLE SYSTEMS

Systems of many particles present some peculiarities which may all be illustrated by considering a two-particle system. What we expect, and what is indeed the case, is that the quantum states for the two-particle system are obtained in some way from those for the two one-particle systems. First consider the case where the two particles are of different kinds (with different masses, say). Then if u is a possible state for the first particle and v a possible state for the second, there is a possible state $u \times v$ for the two-particle system. This is interpreted as the first system being in u and the second in v. The states of two-particle systems obtained in this way I shall call *decomposable* states.[22]

Now suppose we perform an observation A on the first particle, and an observation B on the second. If the state is decomposable, then we have the gratifying result that the probability of a given result for B is quite independent of whether or not A has been performed.[23] Observing either the position or the momentum of the first particle, for instance, in no way affects the probability for various results of observing the momentum of the second. This result is what we would expect. Observing one particle should not, we think, affect the second, especially if the particles are well separated in space.

However, quantum theory reveals more of the Divine sense of humour than its classical mechanical approximation, and things are not as simple as this suggests. For not all states are decomposable (i.e. of the form $u \times v$). Some are superpositions of decomposable states. And for non-decomposable states we shall not, in general, have probabilistic independence for observations.[24]

It is also worth noting that the decomposable states of, say, a hydrogen atom can evolve into non-decomposable states.[25] So the Schroedinger dynamics can affect the independence of observations.[26]

I now consider two particles of the same kind. In that case it is no longer possible to observe which particle is which. That is, the states $u \times v$ and $v \times u$ would be observationally equivalent in the sense that they would lead to exactly the same probabilities for observations. A typical YNO would be whether there is one particle in region U and another in region V, where it is not specified which particle is in which region. Such YNOs are symmetrical.

It is worth considering just how radical this symmetry requirement is. What happens in the case of classical mechanics? Well, we can point to *this* particle and to *that* one and then observe whether *this* is in region U and *that* in region V. But if the particles swapped positions and other characteristics quickly enough, then we would be none the wiser. For *this* particle is pointed out simply as the one *here* and *that* as the one *there*. We cannot, even in the classical case, observe which particle is which, unless indeed *we* are the particles.

In classical mechanics we can use the observation of position to label particles and *then* distinguish them for the purpose of other measurements. But the same holds in quantum theory. Suppose I have observed that there are two particles, one 2 cm above a horizontal plane, the other 2 cm below it. Then I can distinguish observing the horizontal components of momentum for the particle above the plane from observing those components for the particle below the plane.[27] All that quantum theory prevents in this case, but classical mechanics allows, is the observation of the vertical component of the momentum for the two distinguished particles. For that could so disturb them as to make them swap over, without my knowing.

It is not the unobservability of swapping which is novel, then. But there is a novelty. Suppose for simplicity that u and v are orthogonal states for a single particle of the kind in question. Then $u \times v$ and $v \times u$ would be orthogonal. We can consider superpositions of them. Now consider the observation of one component of the momentum of one particle and, say, the same component of the position of the other particle. In a decomposable state $u \times v$ or $v \times u$ we would have independence, that is, the result of making one of these observations would not affect the probability of making the other. But for a superposition of these we have dependence. While, then, there is no observational difference between $u \times v$ and $v \times u$, these can be distinguished, given repeated experiments, from a superposition of them. And quantum theory tells us that in fact only a superposition occurs, namely a state, call it $u \bigstar v$, which is half way between $u \times v$ and $v \times u$.[28]

It is this lack of decomposable states which is responsible for the special character of the quantum theory of two particles of the same kind. For there are no states that guarantee probabilistic independence of observations on the different particles.

The other interesting feature is the division of all particles into fermions and bosons. Bosons, which include mesons and photons, are the particles of integer spin or no spin at all. Fermions include electrons, protons and neutrons, and are the particles of spin half or some integer plus a half. Bosons and fermions do not differ significantly in the case where u and v are orthogonal. But at the other extreme where $u = v$ they differ in that for bosons but not for fermions there is a state $u \bigstar u$ corresponding to the two particles being in the same state.

V MIXED STATES AND COLLECTIONS OF SYSTEMS

The states I have been considering, even superpositions, are *pure* states. But there are other states called *mixed* states. Given any two states

u and v and any real number λ from 0 to 1 we can form a *mixture* of u and v in the proportions λ to $(1 - \lambda)$. I call this mixed state $\lambda u + (1 - \lambda)v$. By mixing states which are already mixed we obtain mixtures of any finite number of pure states.[29]

Mixture is not necessarily a metaphor. In the quantum version of statistical mechanics we consider large collections of systems. Indeed it seems traditional to idealize the situation by considering *infinite* collections. However, given the problems of defining proportions in infinite collections, I shall concentrate on large but finite collections of systems. (I shall refrain from calling these large collections *ensembles* because that term is sometimes restricted to the case of infinite collections.)

In quantum statistical mechanics pure states describe not only individual systems but also *suitable* large collections of quantum systems, namely those in which the individual systems are all in the same pure state.[30] In that case, we can literally mix collections in pure states to obtain new, larger, collections in mixed states. For example, if we mix 30,000 systems forming a collection described by pure state u and 70,000 systems forming a collection described by pure state v, we obtain a collection of 100,000 systems described by the mixed state $(0.3)u + (0.7)v$. The Statistical Interpretation of Mixture is the thesis that *all* mixed states arise in this way.

However, *mixture* need not be taken literally. One metaphorical use of the word 'mix' concerns our ignorance. If we do not know whether a given system is in state u or state v, then we can think of it as a member of a possible collection of systems, some in state u and some in state v. The proportions in which we 'mix' these possible systems will reflect our relative confidence that the system is in u or in v. (Such a collection of possible systems is often called a *virtual* ensemble.) The Ignorance Interpretation of Mixture, while allowing that mixed states can describe literal mixtures, insists that often they describe our own ignorance.

There is another interpretation of mixture, which I call the Sui Generis Interpretation. We may insist that individual systems can be in mixed states, quite independent of our ignorance. In this case mixture is a basic and unanalysable operation on states. And:

(1) The pure states are merely those states which cannot be expressed as mixtures of other states. They form an interesting but not especially fundamental class of states.

(2) Mixture is largely metaphorical. For, outside of quantum statistical mechanics, mixed states are not literally obtained by mixing collections.

(3) There will be collections of different ('hidden') states which are observationally indistinguishable and which correspond to the same mixed state. For instance, a collection in which 100 per cent of the systems are in the mixed state $1/2u + 1/2v$ will be indistinguishable from a literal mixture of 50 per cent of systems in state u and 50 per cent in state v.

In chapter 3 we shall discuss which speculation is the most reasonable.

The difference between mixture and superposition should be emphasized.[31] Suppose u and v are orthogonal pure states. Then it is easy to show that no superposition of u and v could possibly be the same as any non-trivial mixture of u and v. For the mixture of two different pure states is never a pure state. But a superposition of them is always a pure state.[32]

Having introduced large collections of systems in order to explain mixture, it is convenient at this point to say something about the connection between von Neumann dynamics and these large collections. For a given YNO A, there will be eigenstates of A and eigenstates of $\bar{A}$. The von Neumann dynamics specifies the probability that a given pure state u will change (under observation) to an eigenstate for A and the probability that it will change to an eigenstate for $\bar{A}$. (These probabilities add up to one.) It also specifies which eigenstates would result. This I take to be part of the *formalism* of quantum theory, rather than a speculation, which is why I did not call it the Projection *Postulate*. This, however, is a convenient place to distinguish the so-called Weak and Strong Projection Postulates.

First, what implications does the von Neumann dynamics have for large collections? A mixture of sub-collections in pure states will change into the corresponding mixture of the sub-collections resulting from the von Neumann dynamics for each pure state. So we may concentrate on large collections described by pure states.

Each system is in the same pure state u. So for each there is the same probability, μ, say, of a transition to a specified eigenstate, v, for the YNO A. There will be a probability $1 - \mu$ of transition to some specified eigenstate w for $\bar{A}$.

The proportion of systems which go to v is very likely to approximate the probability μ, of a transition to v. And similarly for the transition to w. So, it is likely that the collection itself will come to be in a mixed state, approximating $\mu v + (1 - \mu)w$. The *Weak Projection Postulate* is based on this. It says that the collection will change from a pure state u to the mixed state $\mu v + (1 - \mu)w$. This amounts to an idealization. We assume something occurs when in fact it is merely highly probable that it occurs to a good approximation. We have already made one substantial idealization in considering von Neumann dynamics. For it requires instantaneous interaction with the system. And the further idealization of the Weak Projection Postulate is harmless enough.

Suppose we have a large collection of systems and make a YNO. Then the resulting collection is specified statistically by the Weak Projection Postulate. However, we can still consider the individual systems observed. The Strong Projection Postulate tells us that the individual systems go into eigenstates for A or for $\bar{A}$ with the probabilities indicated by the theory. So it is just the von Neumann dynamics. The point of calling this the *Strong* Projection Postulate is to contrast it with the *Weak* Projection Postulate. For some purposes we might think of observation as a two-stage process. In the first stage the observation affects the collection of systems in accordance with the Weak Projection Postulate. In the second, one system is picked out from the large collection. The combination, namely the von Neumann dynamics, is then called the Strong Projection Postulate.

VI POSITION AND MOMENTUM

The idealized observations for which von Neumann dynamics holds do not cover everything we would like to observe. For instance, consider YNOs A and B, where A is the YNO of whether the particle is in a specified region V, and B is whether its momentum is in a specified set W of momenta. We would have liked there to be a YNO with a Yes result just in case the results of both A and B are Yes. But the theory does not permit us to assume there is such an observable of the idealized sort being considered. There is nothing to stop us recording these results; we perform A first and then B. Let us call the 'YNO' so obtained $A \circ B$. Then the theory tells us that $A \circ B \neq B \circ A$. The order matters.[33] In that case A and B are said *not to commute*. And neither $A \circ B$ nor $B \circ A$ is one of the observables for which von Neumann dynamics holds, for neither is repeatable. But suppose for the moment that Nature had been more forthcoming. Suppose $A \circ B$ had been the same as $B \circ A$ and had been a genuine YNO. Then for a given state *s*, for any region *V* of space, and for any set *W* of momenta, we would have had a probability $\Pr(V, W)$ that the result of discovering both whether the particle is in region *V* and whether it has momentum in *W* is Yes. As it is, this is not well defined. For, as I have said, $A \circ B$ is not the same as $B \circ A$.

There is, then, no joint probability distribution for the results of *observing* position and momentum. None the less it is important to speculate whether there is a joint probability distribution for position and momentum, of some other kind. Why is it important? Because the probability distributions for position alone and for momentum alone fail (even jointly) to specify the state of a single spinless particle (Corbett and Hurst, 1978). It is a plausible conjecture that a joint distribution would be enough to specify the state. If this conjecture is correct, then we would have analysed – and so understood – the quantum state in simpler terms as the joint distribution for position and momentum (and spin). This would make quantum theory *somewhat* like classical mechanics.

Quite how such a joint distribution would be interpreted need not concern us here. But we might expect that if *W* is the set of all momenta, $\Pr(V, W)$ would be $P_L(V)$, the probability that the result of measuring location is in *V*.[34] Likewise if *V* is the whole of space, $\Pr(V, W)$ would be $P_M(W)$, the probability that the result of measuring the momentum is in *W*. Suppose, then, that to each state we can associate an assignment of joint probabilities $\Pr(V, W)$ satisfying the above constraints. Suppose also that they satisfy a natural linearity condition, namely that if $\Pr(V, W)$ and $\Pr^*(V, W)$ are associated with s and s* respectively, then $\lambda \Pr(V, W) + (1 - \lambda) \Pr^*(V, W)$ is associated with the mixed state $\lambda s + (1 - \lambda)s^*$. Next, suppose that, like ordinary probabilities, the $\Pr(V, W)$ being considered are never negative. Finally, suppose that there is a unique assignment of joint probabilities for each quantum state. Then as Wigner has shown[35] these suppositions jointly violate the standard methods of representing states. (Not surprisingly Wigner's proof makes critical use of superpositions.)

The most obvious premiss to question is the uniqueness of the assignment of joint probabilities. Perhaps every quantum state is a mixture of hidden states each associated with its own joint probability distribution. However, if these mixtures satisfy the linearity condition mentioned, then Wigner's argument applies to the hidden states also.

In spite of Wigner's result we can persist in treating the quantum state as specified by the joint distribution, by allowing negative probabilities. (I shall defer until chapters 4 and 5 the question of how negative probabilities are possible.) We may then rely on Wigner's simple formula[36] for this distribution, and on his equation for the Schroedinger dynamics.[37] As we conjectured, the joint probability distribution does specify the quantum state.

We may also relate the abstract Hilbert-space theoretic structure of the states to the joint probability distributions.[38]

VII SO-CALLED QUANTUM LOGIC[39]

Much recent work on the foundations of quantum theory puts the emphasis on possible observations, in particular on YNOs,[40] and characterizes the states in terms of the observables. I shall complete this chapter by sketching this approach. This section could be omitted on a first reading.

There are various speculations which could motivate this approach, and which, in turn, derive *some* measure of support from the mathematical elegance of the approach. This support should not, however, be exaggerated. The considerations I shall provide against the speculations in question outweigh any derived from mathematical elegance. One of the speculations that would motivate this approach is that, indeed, quantum theory requires a new logic. It is convenient to argue against this at the end of this section. The others are the Kantian and semi-Kantian speculations.[41] In both of these the states are characterized in terms of YNOs.

We begin by considering various relations between YNOs. I shall take the YNOs not to be particular observations, however, but types of observation. Otherwise we could never *repeat* a YNO. Moreover, I shall take two particular Yes/No observations to be instances of the same YNO if their *standard* verbal expressions are logically equivalent.[42] Suppose that one particular observation is the measurement of whether the particle (in a one-particle system) has a characteristic standardly expressed as 'being X'. And suppose that the other is the measurement of whether the particle has a characteristic standardly expressed as 'being Y'. Then if 'being X'

entails 'being Y' and vice versa these measurements are instances of the same YNO.

We have, then, a set of YNOs for a given system. It will not contain all Yes/No observations that could be coherently described, only those for which the von Neumann dynamics holds. So we shall have YNOs about whether the particle is in region V or whether it has momentum in W, but not the YNO of whether it has both position in V and momentum in W.

The YNOs form a partially ordered set, where the ordering is that of entailment. YNO A entails YNO B ($A \Rightarrow B$) if a Yes answer for A entails a Yes answer for B when they are asked of the same system.[43] Likewise we have an operation of negation (or equivalently a relation of being contradictories): $\bar{A}$, the negation of A, is just the YNO which has a Yes answer when A has a No answer and vice versa. The classical rules govern entailment and negation. Thus, if $A \Rightarrow \bar{B}$, $B \Rightarrow \bar{A}$ and $A \Leftrightarrow \bar{\bar{A}}$. As a consequence, $A \Rightarrow \bar{B}$ and $B \Rightarrow \bar{A}$ are equivalent, and we may say that in either case A and B are contraries (or in the jargon, A is *orthogonal* to B).

If we rely on the usual representation[44] we may follow Birkhoff and von Neumann (Hooker, 1979, pp. 1–26) and take this partially ordered set to be a lattice. That is, for given YNOs A and B there are unique YNOs $A \wedge B$ (the meet) and $A \vee B$ (the join) characterized as follows:

(ia) $A \wedge B \Rightarrow A$, and $A \wedge B \Rightarrow B$
(ib) If $Z \Rightarrow A$ and $Z \Rightarrow B$ for some YNO Z, then $Z \Rightarrow A \wedge B$
(iia) $A \Rightarrow A \vee B$ and $B \Rightarrow A \vee B$
(iib) If $A \Rightarrow Z$ and $B \Rightarrow Z$ for some YNO Z, then $A \vee B \Rightarrow Z$.

This characterization could tempt us to assume that meet is conjunction and join disjunction. Resist the temptation at least until this has been discussed further.

It is easy to check that $(\overline{A \wedge B}) = (\bar{A} \vee \bar{B})$ and $(\overline{A \vee B}) = \bar{A} \wedge \bar{B}$. Now consider the distributivity rules, namely:

$$A \wedge (B \vee C) = (A \wedge B) \vee (A \wedge C) \text{ for all YNOs } A,\ B,\ C$$

and:

$$A \vee (B \wedge C) = (A \vee B) \wedge (A \vee C) \text{ for all YNOs } A,\ B,\ C.$$

Since $(\overline{A \wedge B}) = \overline{A} \vee \overline{B}$ etc., one distributivity rule holds if and only if the other holds. So we may concentrate on:

$$A \wedge (B \vee C) = (A \wedge B) \vee (A \wedge C).$$

If it holds, then the lattice in Boolean. But it does not. Instead of $A \wedge (B \vee C) = (A \wedge B) \vee (A \wedge C)$ we have the weaker *Orthomodular Principle*:

If A and B are any YNOs such that $A \Rightarrow B$ then $B \Leftrightarrow A \vee (B \wedge \overline{A})$.

This is a *special case* of distributivity.

Although it is not essential to this formulation of quantum theory, it is helpful, at least as a heuristic device, to represent the YNOs geometrically (as subspaces of a Hilbert space).[45]

I shall digress for a while and explain how this is done. For every YNO A, there is an eigenspace Eig(A). This is a set of states. If it contains only a single member, then A is represented by the ray (i.e. one-dimensional subspace) representing that state. If it contains more than one member, then the rays representing the state will fill up a subspace of higher dimension. For instance, Eig(A) might consist of all the superpositions of two orthogonal states u and v. In that case, A is represented by a plane (i.e. a two-dimensional subspace) which contains the two rays representing u and v. For any YNO X let $\Sigma(X)$ denote the subspace assigned to X. Then $\Sigma(A) = \Sigma(B)$ just in case Eig(A) = Eig(B). In that case, A and B are observationally indistinguishable and so we could speculate that $A = B$.[46] Hence, we would have a one-to-one correspondence between the YNOs and some of the subspaces.[47] Standardly it is assumed that there is such a correspondence and that, moreover, *all* the subspaces are obtained in this way.

There are two trivial YNOs. The one whose answer is always Yes, call it Y, has every state as an eigenstate. So $\Sigma(Y)$ is the whole space. The trivial YNO whose answer is always No, call it N, has no eigenstates. We stipulate that $\Sigma(N)$ is the set containing only the null vector.

Given the standard assumptions relations between YNOs correspond to relations between subspaces as follows:

(a) $A \Rightarrow B$ just in case $\Sigma(A) \subseteq \Sigma(B)$[48]
(b) $\Sigma(\overline{A}) = [\Sigma(A)]^{\perp}$, the set of all vectors orthogonal to every vector in $\Sigma(A)$.

From (a) we have:

(c) $\Sigma(A \wedge B) = \Sigma(A) \cap \Sigma(B)$

and:

(d) $\Sigma(A \vee B) = \Sigma(A) \oplus \Sigma(B)$ the smallest subspace containing both $\Sigma(A)$ and $\Sigma(B)$.

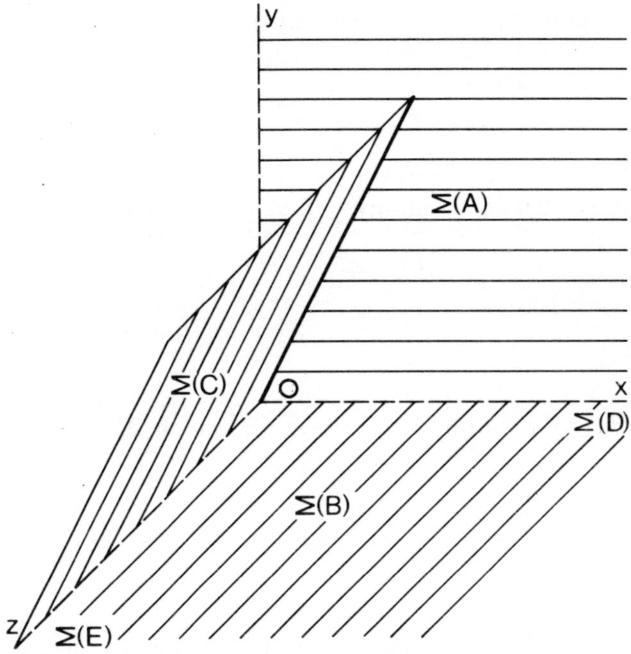

Figure 2.5 Subspaces representing YNOs *A B C D* and *E*.

We may illustrate this using our three-dimensional analogue (see figure 2.5). Let *O* be the origin and *Ox*, *Oy* and *Oz* be three orthogonal axes. Consider the YNOs *A*, *B*, *C*, *D* and *E* where:

$\Sigma(A)$ is the *Oxy* plane.
$\Sigma(B)$ is the *Oxz* plane.
$\Sigma(C)$ is a plane containing *Oz* but making 45° with *Ox* and *Oy*.
$\Sigma(D)$ is the *x* axis.
$\Sigma(E)$ is the *z* axis.

Then $\Sigma(E) \subseteq \Sigma(C)$. Therefore by (a) above we have:

(e) $E \Rightarrow C$.

$[\Sigma(E)]^{\perp}$ consists of all the vectors orthogonal to the z axis. Hence $[\Sigma(E)]^{\perp}$ is the *Oxy* plane, $\Sigma(A)$.

Therefore by (b) above:

(f) $\bar{E} = A$.

$\Sigma(A) \cap \Sigma(B)$ consists of the x axis, $\Sigma(D)$.

Hence by (c) above:

(g) $A \wedge B = D$.

The smallest subspace containing $\Sigma(D)$ and $\Sigma(E)$ is the *Oxz* plane, $\Sigma(B)$.

Hence by (d) above:

(h) $D \vee E = B$.

Now $\Sigma(D) \cap \Sigma(E)$ contains only the null vector, $\Sigma(N)$.

So by (c):

(i) $D \wedge E = N$.

In this case, D and E are orthogonal YNOs, that is $D \Leftrightarrow E^{\perp}$. So if we take $\wedge$ to be like conjunction, result (i) is like an instance of the law of non-contradiction. However, we can have YNOs which are not orthogonal but whose meet is N. This should already make us hesitate before we think of the meet as conjunction. For example $\Sigma(A \wedge C)$ is a line in the *Oxy* plane making 45° with *Ox* and *Oy*. So while $\Sigma(D)$ and $\Sigma(A \wedge C)$ are not orthogonal, their intersection is just the null vector.

Hence:

(j) $D \wedge (A \wedge C) = N$.

This departure from the classical situation is due to the occurrence of YNOs represented by subspaces (lines and planes) which intersect neither at 0° nor at 90°. Such pairs of YNOs are said to be *non-commuting*.[49] If all YNOs were pairwise commuting, then we would have the classical-type result that $A \wedge B = N$ only if $A \Rightarrow B^{\perp}$. We would also have distributivity. Indeed the lattice would be Boolean and the 'logic' entirely classical. The failure of distributivity may be illustrated by considering four YNOs D, E, F, G, all represented by lines in the Oxz plane (see figure 2.6).

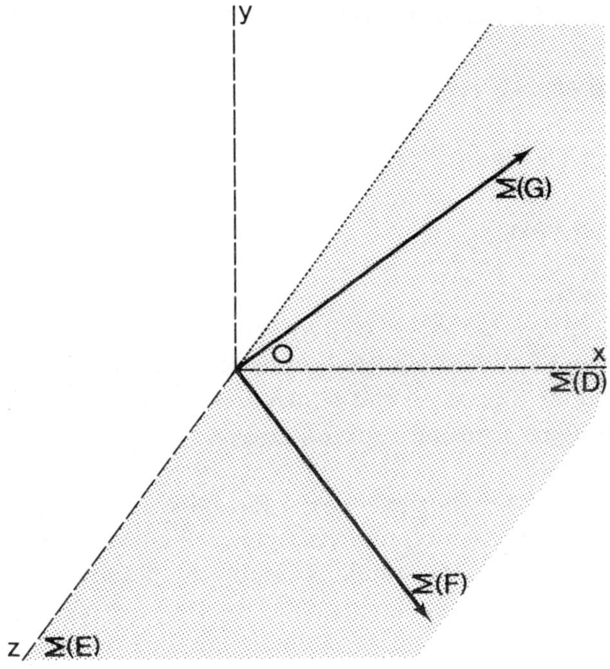

Figure 2.6 Rays, $\Sigma(G)$ and $\Sigma(F)$ in the same plane as $\Sigma(D)$ and $\Sigma(E)$.

$\Sigma(D)$ is the x axis and $\Sigma(E)$ the z axis as before. $\Sigma(F)$ and $\Sigma(G)$ both make angles of 45° with the x and z axes. They are in the Oxz plane and are orthogonal. The smallest space containing both $\Sigma(D)$ and $\Sigma(E)$ is the Oxz plane. So $\Sigma(F) = \Sigma(F) \cap \Sigma(D \vee E)$.

Hence:

(k) $F = F \wedge (D \vee E)$.

But $\Sigma(F) \cap \Sigma(D)$ and $\Sigma(F) \cap \Sigma(E)$ both consist only of the null vector.

So:

(1) $(F \wedge D) \vee (F \wedge E) = N \vee N = N.$

From (k) and (l) we see that distributivity fails.

The Orthomodular Principle is illustrated using F and B (figures 2.5 and 2.6). We have $F \Rightarrow B$ and so need to check that $B = F \vee (B \wedge \bar{F})$. $\Sigma(\bar{F}) = \Sigma(F)^{\perp}$, the plane containing the y axis and $\Sigma(G)$. Hence $\Sigma(B \wedge \bar{F}) = \Sigma(G)$. Therefore:

(m) $F \wedge (B \vee \bar{F}) = F \vee G.$

Now $\Sigma(F \vee G)$ is the smallest space containing $\Sigma(F)$ and $\Sigma(G)$, namely $\Sigma(B)$.

Hence we have:

(n) $B = F \vee G.$

From (m) and (n) we have the required result that $B = F \vee (B \wedge \bar{F})$.

I now end the digression and return to our discussion of the YNOs, considered apart from any representation.

As long as I am merely *expounding* a formalism in which the YNOs are taken as fundamental and the states are characterized in terms of them, I shall retain the standard assumption that we have, indeed, a lattice. Later, when I argue against treating the partially ordered set of YNOs as a logic, I shall relax this assumption, in order to provide a more general result.

The partially ordered set of YNOs is the so-called quantum logic. I say 'so-called', but at this stage that is unfair. However, I can say even at the outset that it is not *obviously* a logic. For the operations of join and meet are not obviously purely logical. I can illustrate this by means of an example. Suppose we were to consider a 'logic' for statements made by Plato. Some entail others. Some are the contradictories of others and so on. Now what will be the meet, $p \wedge q$, of any two such statements p and q? It is the weakest statement Plato ever made which entails them both. If he ever stated the conjunction p and q then the meet would indeed be the conjunction. But if he did not it might be a stronger statement. And similarly the join $p \vee q$ will be the inclusive disjunction, namely the statement that p and/or q, provided Plato at some time stated the disjunction. But if he did not, then it might be some weaker statement. So the meet and join are not determined by purely logical considerations, but also by the extra-logical consideration of what statements Plato made. Likewise the meet and join of the two YNOs are determined not just by obviously logical relations but also by the not obviously logical matter of what YNOs there are, that is of what possible Yes/No observations satisfy the von Neumann dynamics. We might interpret this possibility as logical possibility. In that case the meet and join are determined by logical considerations in the broadest sense and 'quantum logic' would not be a misnomer. But we are also free to treat this possibility as a merely nomological or physical possibility, namely whatever is in accordance with the laws of nature. In that case the so-called quantum logic would be a matter of logic-cum-physics.

Whether this is a logic or a 'logic' we still have a partially ordered set of YNOs. But where do *states* come in? A state s must assign to a YNO A a probability from 0 to 1, namely the probability of a Yes answer. Let us write that probability as $s(A)$. It can be shown that for any YNOs, A and B:

(i) If $A \Rightarrow B$, $s(A)$ is no greater than $s(B)$.
(ii) $s(A \vee \bar{A}) = 1$, and $s(A \wedge \bar{A}) = 0$.
(iii) If A and B are orthogonal (i.e. contraries), then $s(A \vee B) = s(A) + s(B)$.

We can use this to characterize a state as any assignment of probabilities to YNOs with the required properties (i), (ii) and (iii). This amounts to saying that every suitable assignment arises from some state and that no two states give the same assignment. It is a consequence of a result of Gleason's (Piron, 1976, p. 86) that the states characterized in this way will be precisely the assignments of probability given by pure or mixed states in the other formalism. Mixture has a simple interpretation. If s is a mixture of u and v in proportions λ to $(1 - \lambda)$, then for all YNOs A, $s(A) = \lambda u(A) + (1 - \lambda)v(A)$.

The Schroedinger dynamics is simply a matter of the evolution of states, considered as assignments of probabilities to YNOs. Now consider the von Neumann dynamics. The eigenstates for a YNO A are just those states s such that $s(A) = 1$. Now if s is a pure state, that is, one which cannot be expressed as a mixture of other states, then by the above-mentioned result of Gleason's, there is an associated YNO, the *support* of s (Beltrametti and Cassinelli, 1981, p. 117), which I write as E_s. It is the meet of all the YNOs for which s is an eigenstate. That is:

(a) If $s(A) = 1$ then $A \Rightarrow E_s$
(b) Suppose for some YNO E that whenever $s(A) = 1$ then $A \Rightarrow E$. In that case $E_s \Rightarrow E$.

To characterize the von Neumann dynamics we need two items. The first is the probability of a Yes result for the YNO A. If the state is s, this is just $s(A)$. The second is the effect of a Yes result. This will be to change the state from s to an eigenstate u for A. But which eigenstate? The answer is given by the Sasaki projection (Beltrametti and Cassinelli, 1981, p. 129).

We have:

$$E_u = (E_s \vee \bar{A}) \wedge A.$$

The lattice-theoretic approach, then, gives us a satisfactory alternative way of formulating quantum theory. It only remains to discuss in greater detail whether we have a non-standard logic or merely a 'logic'. Clearly if we interpret the meet as conjunction, and the join as disjunction, then since the lattice is not Boolean we have a quite radical departure from standard logic. But as my example of the 'logic' of Plato's statement shows, the meet could well be stronger than conjunction and the join weaker than disjunction. What we can say is that when there is a YNO which is a conjunction then it is a meet, and when there is a YNO which is a disjunction then it is a join. But I shall argue that, in general, we cannot expect there to be conjunctions and disjunctions of YNOs. However, we do have conjunction and disjunction whenever the results of the YNOs A and B can both be *calculated* without further measurement from a single (idealized) observation. Thus we can measure the location of a particle and from the result calculate the answer to the YNOs of whether it is in U and whether it is in V. In such cases we can also calculate the answer to the conjunction of the YNOs and to their disjunction. So we do have conjunction and disjunction. And where we have them, as I said, they must be the meet and join.

A trivial case in which there is a disjunction and a conjunction is that in which YNO A entails YNO B. Then the conjunction just is A again, and the disjunction is B again. A non-trivial case is that in which A and B are orthogonal (i.e. contraries). Then the disjunction is not trivial. For example, suppose A is the YNO of whether the particle is in region U and B the YNO of whether it is in region V. Then if U and V are disjoint, A and B are orthogonal. In that case the disjunction is whether the particle is in $U \cup V$, the union of the two regions.

Where there are conjunctions and disjunctions, then, these will be meets and joins, respectively. But I claim there are often no conjunctions and no disjunctions. If I am right, then the quantum 'logic' is compatible with classical logic, but has a different subject matter. You may say: Why cannot we simply define the conjunction of A and B as the meet? The answer is that we are looking for a conjunction which is a YNO, and what YNOs there are (that is idealized Yes/No observations for which von Neumann dynamics holds) is a matter of scientific discovery and not stipulation. Furthermore, if we try defining the conjunction of A and B, how will we do it? We might consider the observation $A \circ B$, in which we first observe whether A then whether B and record a Yes answer to $A \circ B$ if and only if we have a Yes answer for both A and B. But two difficulties arise. First $A \circ B$ will not, in general, be the same as $B \circ A$. Second $A \circ B$, although a Yes/No observation of sorts, is not one of those for which von Neumann dynamics holds, for it is not repeatable. Both these results follow from the details of the formalism in the case where A is the measurement of whether the particle is in some finite region V and B is whether it has momentum in a set W of finite diameter. And if we wonder how it is possible that $A \circ B \neq B \circ A$, we can see that it is possible because we could think of measuring A as inevitably disturbing the state in such a way that the probabilities for the measurement of B are altered. Indeed, as Heisenberg showed (1930, pp. 21–5), a model for such inevitable disturbance could be based on classical mechanics: measuring position very accurately inevitably disturbs momentum and vice versa.

$A \circ B$, then, is not a YNO, and even if we insisted on adjoining $A \circ B$ to the partially ordered set of YNOs it would not provide us with a meet of A and B. For if it did, by parity so would $B \circ A$. But the meet is unique and $A \circ B \neq B \circ A$.

I have argued that the attempt to define a conjunction and a disjunction of YNOs will in general fail. But perhaps, you say, it is not a matter of definition so much as hypothesis. We have these meets and joins, and would it not be simpler to identify them with conjunctions and disjunctions? I have an argument against that proposal also. My argument is to show that one of the failures of distributivity required on this hypothesis is so spectacular that we should reject it, even if we are prepared to countenance the failure of distributivity in general.

Let us take a very large but finite region of space V. Likewise let us take a very large set of momenta W of finite diameter. Let A and B be the corresponding YNOs. Even if we do not assume that two YNOs *always* have a meet, in *this* case A and B have a meet which is a YNO N whose answer is always No. For if not, there is some YNO C whose answer is not always No, such that $C \Rightarrow A$ and $C \Rightarrow B$. Let u be an eigenstate for C. Then u is a common eigenstate for A and B. But they have no common eigenstate.[50] This is a *reductio ad absurdum* of the supposition that A and B either fail to have a meet or have a meet other than N.

Hence, if we take meets to be conjunctions, then no matter how large are V and W the answer to whether the particle is in V with momentum in W is No with probability one. It is not, note, that we do not get an answer: it is that the answer is inevitably No. This would lead to a contradiction if we had (infinite) distributivity. But even without distributivity it is still a most peculiar result. To illustrate this peculiarity, suppose we had an enormous vacuum – one as large as the observable universe – and we put in it one electron without doing anything to it which would give it a great velocity. It would ordinarily be taken as a negligible idealization to assume that the electron is in that region and has a momentum less than that for an electron with a velocity of 99 per cent of the speed of light. Suppose we are now told that, contrary to what is often said, there *is* an observation of whether the electron is both in the region and has momentum less than that for one with velocity 99 per cent of the speed of light. We would expect the answer to be Yes with a probability just less than one. But we are being told it has probability zero. If we could show that this result followed from the formalism of quantum theory we would be perplexed: something would have gone wrong somewhere but we might not know where. But in this case it follows from a hypothesis, a conjecture which has no observational consequences, based only on the judgement that it is simpler to say that meets are conjunctions rather than to deny that YNOs always have conjunctions.

I conclude that there are no reasons for taking the meet to be a conjunction and that to call the lattice a logic is a misnomer. In addition I judge that any Putnamesque appeal to failure of distributivity when discussing, say, the twin-slit thought experiment must be judged entirely on its success or failure in removing specific difficulties, and cannot be supported by an appeal to the non-Boolean character of the lattice of YNOs.

NOTES

1 The Copenhagen Interpretation, as it is called, is somewhat nebulous. And I am no historian of ideas. However, this much seems clear. The theory which now goes by the name of quantum mechanics owes its formulation to Schroedinger, on the one hand, and Bohr, Heisenberg, Born, Dirac and von Neumann, all roughly of the Copenhagen School, on the other. See Mehra (1974, pp. 4–5).

2 For introductory texts see Messiah (1961, 1962), Das and Melissinos (1986) and Beltrametti and Cassinelli (1981).

3 This is usually taken to be an infinite-dimensional *separable* complex Hilbert space. The one-dimensional subspaces are in one-to-one correspondence with the orthogonal projection operators of rank one. So the pure states may also be represented by such projection operators. If so, then the mixed states are represented by positive operators of trace one. All this is discussed in von Neumann (1955) and Beltrametti and Cassinelli (1981).

4 The chief disanalogy is that a three-dimensional space can be thought of as a real but not as a complex space.

5 If x and y are any two non-zero vectors in a Hilbert space, the 'angle' between them, θ, is given by the angle (from 0° to 90°) for which: $\langle x|x\rangle \times \langle y|y\rangle(\cos\theta)^2 = |\langle x|y\rangle|^2$. Here $\langle x|y\rangle$ is the inner product defined by the Hilbert space. Then the 'angle' between two pure states is just the 'angle' between any two non-zero vectors in the one-dimensional subspaces representing them. The reader may like to check that this 'angle' has the formal properties of a metric. It is possible to characterize orthogonality and superposition in terms of the 'angle'. (Thus s is a superposition of orthogonal states u and v just in case the 'angle' between u and s plus the 'angle' between s and v equals the 'angle' between u and v, namely 90°.) This suggests that we could have treated the 'angle' as fundamental rather than the Hilbert space structure. However, I shall stick to the standard Hilbert space formalism in these notes.

6 That is, the vectors in the subspace representing s are complex linear combinations of vectors in the subspaces representing u and v.

7 In the three-dimensional analogue, for every angle θ strictly between 0° and 90°, there will be two rays in the plane of R_u and R_v making θ with the ray R_u. But because the Hilbert space is a complex vector space, for every 'angle' θ strictly between 0° and 90° there will be infinitely many superpositions of u and v, not just two. These are in one-to-one correspondence with the points on a circle. Indeed the set of all superpositions of two given orthogonal states u and v has a metric structure isomorphic to a sphere with u and v at the poles. If we measure the distance on the sphere along great circles and measure the angle in radians, then we should assign the sphere a radius 1/2, so that the poles are a distance $\pi/2$ apart.

8 Here I use what is called the *Schroedinger picture*. There is also the *Heisenberg picture* according to which, in the Schroedinger dynamics, the state is unchanged, but there are time-dependent observables.

9 In terms of the Hilbert space formalism, the change of state is given by a one-parameter group of unitary operators. That is, for each real number t, there is a unitary operator U_t on the Hilbert space. We have:

(i) $U_{t+\Delta} = U_t U_\Delta$ for any real Δ.
(ii) $U_o = Id$, the identity map.
(iii) U_t is a linear map, that is, $U_t(\alpha x + \beta y) = \alpha U_t x + \beta U_t y$, for all complex numbers α and β and all vectors x and y.
(iv) $\langle U_t x | U_t y \rangle = \langle x | y \rangle$ for all vectors x and y.

Then if x is any vector in the one-dimensional subspace representing the state of the system at time t, $U_\Delta x$ will be a vector in the one-dimensional subspace representing the state of the system at time $t + \Delta$.

10 To cut a long story short, the alternative is to consider observables whose results are any real number. They are represented by unbounded Hermitian operators with continuous spectra. If we stick to an ordinary Hilbert space over the standard complex numbers (i.e., not augmented with infinities and infinitesimals), these operators will not have any eigen-vectors, but only approximately-eigen-vectors. This is inconvenient.

11 Throughout this book when I talk of a region V I mean a *measurable* set of locations. Likewise, when I talk of a set of momenta, I mean a *measurable* set of momenta.

12 If Z is any closed subspace of a Hilbert space H, then for every vector x in H, there is a unique pair of vectors u and v such that v is in Z and u is orthogonal to every vector in Z, and such that $x = u + v$. We may define the projection on to Z, I_z as the operator which sends x to v in this way. I_z is linear, it is self-adjoint (i.e., $\langle I_z x | w \rangle = \langle x | I_z w \rangle$ for all vectors x and w), and it is idempotent (i.e., $I_z^2 = I_z$). Conversely, any idempotent self-adjoint linear operator on H is identical to I_z for some closed subspace

Z of H. The projection on to Eig(A) is represented by I_z where Z is the subspace which is the union of all the one-dimensional subspaces representing members of Eig(A). It can be shown that Z is closed.

13 The existence of this minimum is a consequence of the fact that the subspace associated with Eig(A) as in note 12, is *closed*.

14 See note 10. The problem is that the result of an exact measurement should be a state represented by an eigen-vector for a Hermitian operator of continuous spectrum. But, in a standard Hilbert space, there are no such eigen-vectors.

15 Suppose we try to introduce the YNO of whether the momentum is precisely p_o. The wave-mechanical formalism will tell us that its only eigenstate is that given by a sinusoidal wave function:

$$\Phi(q) = k \cos p_o \cdot q + ik \sin p_o \cdot q.$$

Here k is a constant. And $p \cdot q$ is the inner product $p_x q_x + p_y q_y + p_z q_z$. To obtain $\int |\Phi(q)|^2 dq = 1$, we would have to take k to be an infinitesimal. That might be considered problematic, but it does not stop us treating other wave functions as combinations of ones of the form $k \cos p \cdot q + ik \sin p \cdot q$ for various p, and so analysing them into components of precise momentum. This is what the Fourier analysis does.

16 $\hat{\Phi}(p) = c\int \Phi(q) e^{-ip \cdot q}\, dq.$

Here c is a constant chosen so that $\int |\hat{\Phi}(p)|^2\, dp = 1$ when $\int |\Phi(q)|^2 dq = 1$. A calculation shows that $c^2(2\pi)^3 = 1$.

17 The 'angle' was defined in terms of the inner product (see note 5). And the inner product is:

$$\langle \Phi | \Psi \rangle = \int \bar{\Phi}(q) \Psi(q)\, dq.$$

18 For $\alpha\Phi + \beta\Psi$ to represent a superposition of Φ and Ψ, we require that $\int |\alpha\Phi(q) + \beta\Psi(q)|^2\, dq = 1$. If the states are orthogonal, this will be satisfied provided $|\alpha|^2 + |\beta|^2 = 1$.

19 Consider an electron. Its mass is m. Perhaps it has no determinate radius. In that case we can consider its mean radius r. Suppose no part of it moves with velocity greater than the speed of light, c. Then if it had determinate radius x its angular momentum due to rotation would be less than mxc. So the mean angular momentum cannot exceed mrc. Hence if the spin is due to rotation we would expect $mrc > h/4\pi$, where h is Planck's constant. This is not consistent with the values given in standard tables. One possible way round this problem is to posit particles of negative mass. Suppose, for example, that two particles of the same

charge as an electron, but 1,000 times the mass, orbited around a particle with the charge of a proton, and of *negative* mass 1,999 times that of an electron. Then the total mass and charge of the system would be that of an electron, but its rotational angular momentum could be much greater than that of a simple particle.

20 Using the wave formalism (see note 21) the state is represented by a pair of complex functions $\langle \Phi_1, \Phi_2 \rangle$. Without loss of generality, we may take the projection operators corresponding to the two YNOs to be 1/2 $(Id + \sigma_x)$ and 1/2 $(Id + \sigma_y)$ where σ_x and σ_y are Pauli spin matrices. Then the probabilities of a Yes result are λ and μ respectively, where:

(i) $\lambda = \int 1/2(|\Phi_1(q) + \Phi_2(q)|^2 \, dq$.
(ii) $\mu = \int 1/2(|\Phi_1(q) + i\,\Phi_2(q)|^2 \, dq$.

We may consider the functions Φ_1 and Φ_2 to be vectors of a Hilbert space. From (i) and (ii) we have:

(iii) $\lambda = 1/2 + \mathrm{Re}\langle \Phi_1 | \Phi_2 \rangle$

and

(iv) $\mu = 1/2 + \mathrm{Im}\langle \Phi_1 | \Phi_2 \rangle$

Hence:

$$\lambda(1-\lambda) + \mu(1-\mu) = 1/4 - (\mathrm{Re}\langle \Phi_1 | \Phi_2 \rangle)^2 + 1/4 - (\mathrm{Im}\langle \Phi_1 | \Phi_2 \rangle)^2.$$

Hence:

$$\lambda(1-\lambda) + \mu(1-\mu) = 1/2 - |\langle \Phi_1 | \Phi_2 \rangle|^2.$$

By the Cauchy–Schwarz inequality:

$$|\langle \Phi_1 | \Phi_2 \rangle|^2 \leq 1/4(\|\Phi_1\|^2 + \|\Phi_2\|^2) = 1/4.$$

So the sum of the variances is at least 1/4.

21 Suppose the spinor function is $\Phi(q)$, where $\Phi(q)$ is represented by the pair of complex numbers $\langle \Phi_1(q), \Phi_2(q) \rangle$. Then we can represent the state in two other ways. First, we can represent it by an *associated state* for a spinless particle, represented in turn by $\Psi(q)$, and by the spin vector $1/2\,\sigma(q)$. Second, we can represent it by the vector $|\Psi(q)|\sigma(q)$ and by a phase *angle* equal to $\mathrm{Arg}\,\Psi(q)$. $\Psi(q)$ is given by:

(i) $|\Psi(q)|^2 = |\Phi_1(q)|^2 + |\Phi_2(q)|^2$
(ii) $\mathrm{Arg}\,\Psi(q) = \mathrm{Arg}\,\Phi_1(q) + \mathrm{Arg}\,\Phi_2(q)$

The vector $\sigma(q)$ is of unit length.

The x, y and z coordinates of $\sigma(q)$ are proportional to:

(i) $\Phi_1(q) \times \overline{\Phi}_2(q) + \overline{\Phi}_1(q) \times \Phi_2(q)$
(ii) $[\Phi_1(q) \times \overline{\Phi}_2(q) - \overline{\Phi}_1(q) \times \Phi_2(q)]/\mathrm{i}$

and:

(iii) $|\Phi_1(q)|^2 - |\Phi_2(q)|^2$

22 If the first system is represented using the Hilbert space H_1 and the second using the Hilbert space H_2, then the two-particle system can be represented using $H_1 \otimes H_2$, the tensor product of H_1 and H_2. For each member x of H_1 and y of H_2 there is a member $x \otimes y$ of $H_1 \otimes H_2$, called the tensor product of x and y. The inner product on $H_1 \otimes H_2$ is given by the formula: $\langle x \otimes y | w \otimes z \rangle = \langle x|w \rangle \times \langle y|z \rangle$. But not all members of $H_1 \otimes H_2$ are tensor products, for some are (finite or infinite) sums of tensor products. The one-dimensional subspaces whose vectors *are* tensor products are those which represent the decomposable states.

23 A can be thought of as both a YNO for the one-particle and for the two-particle system. Thus if A is the observation of whether the first particle is in region V, it is an observation we can make either on the first particle considered as a one-particle system or on the two-particle system. To avoid confusion, I shall denote the former as $A^{[1]}$ and the latter as $A^{[2]}$. The formalism for composite systems tells us the eigenstates for $A^{[2]}$ are precisely the composable states $u \times v$ such that u is an eigenstate for $A^{[1]}$.

That is:

$$\mathrm{Eig}(A^{[2]}) = \{u \times v : u \in \mathrm{Eig}(A^{[1]})\}$$

Similarly:

$$\mathrm{Eig}(B^{[2]}) = \{u \times v : v \in \mathrm{Eig}(B^{[1]})\}$$

It is then an elementary exercise in the theory of tensor products and Hilbert spaces to show that we have probabilistic independence.

24 Let the projection operators associated with $A^{[1]}$ and $B^{[1]}$ be P and Q respectively. Consider a state represented by a vector of the form $\alpha x \otimes y + \beta w \otimes z$. From the formula, we have probabilistic independence just in case:

$$\|\alpha\, Px \otimes Qy + \beta\, Pw \otimes Qz\|^2 \times \|\alpha x \otimes y + \beta w \otimes z\|^2 =$$
$$\|\alpha\, Px \otimes y + \beta\, Pw \otimes z\|\alpha x \otimes Qy + \beta w \otimes Qz\|^2$$

Now consider the case where $|\alpha|^2 = |\beta|^2 = 1/2$, where x, y, w and z are pairwise orthogonal unit vectors, and where $Px = O = Qy$, but $Pw = w$ and $Qz = z$. Then it is easy to check that the R.H.S. of the above equation is 1/2 and the L.H.S. is 1/4. So we do not have probabilistic independence.

25 Ignoring spin, the states of the two-particle system are represented by functions $\Phi(q, r)$ of two position variables, q and r with coordinates x_1, x_2, x_3 and x_4, x_5, x_6.

The Schroedinger equation for the hydrogen atom is:

$$i\dot{\Phi} = -(1/2m_1)(\Phi_{,11} + \Phi_{,22} + \Phi_{,33})$$
$$-(1/2m_2)(\Phi_{,44} + \Phi_{,55} + \Phi_{,66})$$
$$+ V(x_1, x_2, x_3, x_4, x_5, x_6)\Phi$$

Here m_1 and m_2 are the masses of the proton and the electron. $\dot{\Phi}$ is the first derivative with respect to time, and $\Phi_{,nn}$ is the second derivative with respect to x_n. V is the potential energy. If decomposable states evolved into decomposable states we would have for all times that:

$$\Phi(q, r) = F(q) \times G(r).$$

Where these functions are non-zero, we obtain from Schroedinger's equation:

$$i(\dot{F}/F + \dot{G}/G) = -(1/2m_2)(F_{,11}/F + F_{,22}/F + F_{,33}/F)$$
$$-(1/2m_2)(G_{,44}/G + G_{,55}/G + G_{,66}/G) + V.$$

Of these terms, $\dot{F}/F$; $F_{,11}/F$; $F_{,22}/F$; and $F_{,33}/F$ are functions of q only. And $\dot{G}/G$; $G_{,44}/G$; $G_{,55}/G$; and $G_{,66}/G$ are functions of r only. So we can guarantee an evolution of decomposable states into decomposable states only if the potential $V(q, r)$ is of the form $g(q) + h(r)$, the sum of a function of q and a function of r. But the potential due to electrostatic attraction is not of that form. So decomposable states will evolve into non-decomposable ones.

26 Strictly, all I have shown is that *some* decomposable states evolve into non-decomposable states and that *some* non-decomposable states result in a failure of probabilistic independence. This leaves open the possibility that the states of a two-particle system for which we have probabilistic independence never evolve into ones which do not. In that case, we might insist that all the physically possible states are of this sort. This would be a superselection rule. It is possible to check whether this is so by calculating the wave function for a hydrogen atom and showing that it does not give probabilistic independence. My excuse for not presenting

the calculation is that for systems of two particles of the same kind, it is easy to show we do not have probabilistic independence. See for instance Beltrametti and Cassinelli (1981, sections 7.3, 7.4). Given this, it would be ad hoc, even if it were possible, to propose the required superselection rule for particles of different kinds.

27 The particles are distinguished by a position observation at right-angles to both these components.

28 If u and v are represented by unit vectors x and y, then $u \bigstar v = v \bigstar u$ is represented either by $(1/\sqrt{2})(x \otimes y + y \otimes x)$ or $(1/\sqrt{2})(x \otimes y - y \otimes x)$, depending on whether the particles are bosons or fermions. In either case, the 'angle' this vector makes with $x \otimes y$ is 45° as is the 'angle' it makes with $y \otimes x$. In that sense, $u \bigstar v$ is half way between $u \times v$ and $v \times u$.

29 We may also consider suitable infinite mixtures. Suppose the states u_1, u_2 etc. form an infinite sequence of pairwise orthogonal states. Then, as in note 3, u_k may be represented by an orthogonal projection operator P_k of rank one, namely the orthogonal projection on to the subspace $\{\lambda u_k : \lambda \in \mathbb{C}\}$. Now suppose that λ_1, λ_2 etc. form an infinite sequence of positive numbers whose sum is one. Then there is an infinite mixture $\lambda_1 u_1 + \lambda_2 u_2 + \ldots$, namely the state represented by the operator $\lambda_1 P_1 + \lambda_2 P_2 + \ldots$

30 There is a complication here if we adopt a hidden variables 'interpretation', according to which a system in a given pure state can be in different underlying hidden states. Then even a *pure* state describes a large collection of systems in those various underlying hidden states. The frequencies for the different hidden states should approximate the probabilities specified by the hidden variables 'interpretation'.

31 In distinguishing between mixture and superposition, we should be clear about the system we are considering. If a two-particle system is in a pure but non-decomposable state then the one-particle systems are in mixed states if they are in any quantum state at all (Beltrametti and Cassinelli, 1981, pp. 66, 67).

32 Represent pure and mixed states as in note 3. Then a non-trivial mixture of two pure states is represented by an operator whose *rank* is two. But the superposition of u and v will be represented by the orthogonal projection on to that superposition. This has rank one.

33 For instance, suppose A is the position YNO of whether the particle is in region V and B the momentum YNO of whether its momentum is in W. And consider a system in an eigenstate u for A. Then the probability of ⟨No, No⟩ for $A \circ B$ will be zero only if either u is also an eigenstate for B or the effect of obtaining a Yes result for B turns u into an eigenstate for A. In the latter case, the state obtained is an eigenstate for both B (by repeatibility) and A. In the former case u itself is an eigenstate for both B and A. In either case, then, A and B share a common eigenstate.

As Jauch has remarked (Jauch, 1968, p. 221) and Gibbins and Pearson have proved (Gibbins and Pearson, 1981, p. 803) this is not possible. For suppose there were a common eigenstate and consider the wave function Φ representing it. It must be a function zero outside V. Hence, its Fourier transform is the restriction to the real numbers of an *analytic* function. Hence $\hat{\Phi}$ cannot be zero outside W. But since u would be an eigenstate for B, $\hat{\Phi}(p)$ would have to be zero outside W. This contradiction is a *reductio ad absurdum* of the supposition of a common eigenstate.

34 That is, we expect the probability distribution for the result of observing position to be the same as the probability distribution for position prior to observation. This is the requirement of *Statistical Veridicality*, which I shall discuss in chapter 3. There I argue that Statistical Veridicality is a desideratum for position and for momentum but not for all observables.

35 See 'Quantum-mechanical distribution functions revisited' (Wigner, 1971, pp. 25–36). Wigner assumes that the joint probability distribution is of the form $\langle u | M(q, p)\, u\rangle$ for a pure state u, where $M(q, p)$ is a Hermitian operator depending only on q and p. But he only uses this in one place (equation 4.5 on p. 27). And what he uses is merely that the joint probability is a linear function of the Hermitian operator representing the state. And that is ensured by the linearity constraint which I have assumed, namely (in my notation) that $\lambda \mathrm{P}(v, w) + (1 - \lambda)\, \mathrm{P}^*(v, w)$ is associated with $\lambda s + (1 - \lambda) s^*$. So Wigner has not made any significant assumption other than uniqueness.

36 The joint probabilities are given by a density-function $\mathrm{P}(q, p) = K \int \bar{\Psi}(q + x) \Psi(q - x) \mathrm{e}^{2\mathrm{pix}}\, \mathrm{dx}$. This is based on Wigner's formula (Wigner, 1971, p. 33). I have put Planck's constant equal to 2π. Wigner is considering the one-dimensional case for which $K = (1/\pi)$. In general, I take it $K = (1/\pi)^{3\mathrm{n}}$ if there are n particles.

37 Wigner has a horrendous-looking equation (Wigner, 1932): but it is not as bad as it looks. The equation of motion is:

$$\partial \mathrm{P}/\partial \mathrm{t} + (p/m)\, \mathrm{grad}_{\mathrm{q}} p = \tilde{\mathrm{V}}(\mathrm{P}).$$

Here $\mathrm{grad}_{\mathrm{q}}$ is the vector operator of differentiation with respect to location, and $\tilde{\mathrm{V}}$ is a differential operator involving only derivatives with respect to p.

38 If states u_1 and u_2 are represented by wave functions ψ_1 and ψ_2, then Wigner has a simple formula for $|\langle \psi_1 | \psi_2 \rangle|^2$ (Wigner, 1971, p. 35). Making the same alteration as before we have:

$$|\langle \psi_1 | \psi_2 \rangle|^2 = (1/2\pi)^{3\mathrm{n}} \iint \mathrm{P}_1(q, p) \mathrm{P}_2(q, p)\, \mathrm{d}q\, \mathrm{d}p.$$

39 For a discussion of quantum logic see Beltrametti and Cassinelli (1981). See also the various collections edited by Hooker (1976, 1978, 1979) – especially the papers by Birkhoff and von Neumann (Hooker, 1976, pp. 1–26) and Piron (Hooker, 1976, pp. 513–43). And see Mackey (1963) and Jauch (1968).

40 An alternative is to rely on a C* algebra of observables which are not just Yes/No observables. However we can identify the Yes/No observables as the self-adjoint idempotents (i.e. the element x such that $x^2 = x = x^*$). So I do not think the C*-algebra approach presents any great difference for my purposes. So I ignore it, and ask its devotees to adjust what I say about the 'quantum logic' approach. Likewise I shall not discuss Pool's use of Baer*-semigroups (Pool, 1968a,b) except in these notes.

41 See chapter 3, section 3, and chapter 4, section 5.

42 Some YNOs have *standard* verbal expressions, such as 'whether the particle has location in V', 'whether the particle has momentum in W', 'whether the particle has spin direction up with respect to direction a'. (There will be more complicated standard verbal expressions for systems of several particles.) Moreover, the negation of any standard verbal expression is also a standard verbal expression. So are the conjunctions and disjunctions, although in many cases there will be no YNO with a conjunctive or disjunctive standard verbal expression. Note also that the 'if' does not mean 'if and only if'. We might, for instance, decide to identify a YNO A and a YNO B if A and B are observationally indistinguishable.

43 More precisely $A \Rightarrow B$ if *some* standard verbal expression of A entails *some* standard verbal expression of B. If we insist that observationally indistinguishable YNOs are identical then the same YNO could have more than one standard verbal expression. Note that I have written 'if' not 'if and only if'. Some YNOs might not have standard verbal expressions. In that case the entailment relation is extended to cover those which do not on the grounds of theoretical simplicity. See note 48.

44 In which every YNO A is represented by the union of the one-dimensional subspaces representing its eigenstates. This representation assigns the same subspace to A and B only if $A \Leftrightarrow B$. The subspaces are ordered by inclusion. Equivalently, we may represent the YNOs by the orthogonal projections on to the subspaces. That the partially ordered set is a lattice follows from the non-trivial assumption that every closed linear subspace of the Hilbert space corresponds to a YNO. This is an immediate consequence of Mackey's Axiom VII (Mackey, 1963, p. 71), which, as Mackey notes (Mackey, 1963, p. 72), is not as easily motivated as the other axioms. We could motivate the claim that the partially ordered set is a lattice on the grounds that we are departing as little from the situation in classical mechanics as possible.

In classical mechanics, the observables form a Boolean algebra and hence a lattice. But if we have no further motivation, then the existence of meets and joins would have much the same status as the existence of non-trivial superpositions. There would be merely a defeasible presumption in favour of it.

One attempt to motivate the claim that the partially ordered set is indeed a lattice is based on the work of Pool (1968a) on the semigroup of operations associated with YNOs. (The operation associated with *A* is the one that characterizes the effect on the space of states of a Yes result.)

Pool states various axioms (1968a, p. 127) from which he can show that this is a Baer*-semigroup and that the YNOs form a lattice. However, one of these axioms (II.4) is no easier to motivate than Mackey's Axiom VII. It tells us, for instance, that if the overall effect of a Yes result for YNOs *A* followed by *B* is the same as for *C* followed by *D*, then the overall effect of Yes results for *B* followed by *A* is the same as for *D* followed by *C*. This, like the more general Axiom II.4 itself, has the merit of mathematical elegance. But once again, that establishes no more than a presumption.

I maintain, therefore, that the assumption that the partially ordered set of YNOs is a lattice is not essential to this approach and it could be given up under pressure.

45 As in note 44.

46 This is a matter of some delicacy. By the Radical Privilege, to be discussed in chapter 4, we are entitled so to speculate. But we are not compelled to.

47 See note 44. Here by a *subspace* I mean a subset which is itself a Hilbert space, that is a set closed under linear combination which is also closed with respect to the topology resulting from the Hilbert space structure.

48 This is the extension of the entailment relation mentioned in note 42.

49 Because the corresponding projections do not commute. Commutivity can be defined in lattice-theoretic terms. *A* and *B* commute just in case there are pairwise orthogonal YNOs X, Y, and Z such that $A = X \vee Z$ and $B = Y \vee Z$. See Beltrametti and Cassinelli (1981, p. 125).

50 I cannot justify this claim by deducing it from one of the standard formalisms, as in note 33. For, by considering the possibility that the YNOs do not form a lattice, I implicitly allow some deviation from this formalism. My argument is that, in whatever way we formulate quantum theory, we should preserve a *core* of certain aspects of the theory as standardly developed. This core consists of the results which a working physicist, not especially interested in the foundations of quantum theory, would recognize as a quantum-theoretic result. No doubt this characterization is vague. But some results are definitely not in the core,

while others definitely are. For example, it is not a core result that given any collection of pure states there is a YNO whose eigenspace consists just of the (finite or infinite) superpositions of those states. For the YNO required might well have no simple description in terms of position, momentum and spin. However, whether or not the position YNO A and the momentum YNO B have a common eigenstate would seem to belong to the core. If this is agreed, then, we may *now* infer from note 33 that A and B have no common eigenstate, even if we do not assume the YNOs form a lattice.

3

Puzzles and Problems

In this chapter I introduce the puzzles and problems which make quantum metaphysics both difficult and interesting.[1] Because some of the difficulties are based on thought experiments I should make it clear that a thought experiment is not an attempt to replace actual experiment by a priori considerations. Rather, it is a way of bringing out the consequences of quantum theory. But it so happens that the thought experiments I consider are like actual experiments, and the results are in good agreement with the theoretical predictions. Actual experiments would also bring out the surprising consequences of quantum theory, but not as clearly. For instance the twin-slit experiment cannot be performed with electrons: but very similar phenomena occur with their diffraction by crystals. It is just that the puzzling behaviour of an electron is obscured by the complexities of the theory of diffraction.

I SCHROEDINGER'S CAT

The peculiarities of quantum theory appear chiefly at the microscopic level. But the macroscopic cannot be entirely separated from the microscopic. To bring this out, Schroedinger considered a cat, placed in a box so as to isolate it from observation, whose survival depends on some probabilistic quantum effect. However, since a cat can itself observe its own survival, we do not have the required isolation. Accordingly, in a less cruel thought experiment I consider some cute plaster cats dressed up in various ways, such as Puss in Boots. A.R.M.[2] plans to eliminate them, by placing small soundless bombs, timed to explode at two o'clock in the morning, next to each plaster cat in the neighbourhood. No one will observe the result until dawn. However, A.R.M. are not quite as efficient as most terrorists and, as a result of some quantum phenomenon whose details need not concern us, there is only a 50 per cent chance that a given bomb will explode. Consider a plaster cat. Is it 'alive' or 'dead'? That is, is it intact or scattered in fragments at ten minutes past two? The state of the system is such that there is a 50 per cent probability that the result of an observation of the cat would be that it is intact. Suppose the cat is intact just before it is observed. Then the result of observation could be that the cat is exploded. How odd that the explosion which failed to occur at two o'clock would occur an instant before the person observed the cat, without of course the person noticing the explosion itself! Alternatively, suppose the cat were in a non-intact (i.e. exploded) state just before the observation. Then the result of observation might be an unexploded cat. Are we to say that the cat was reconstituted? Surely not. There is some pressure then to say that, until it is observed, the cat is neither intact nor exploded. But that is peculiar also.

I think there are three basic responses to this difficulty. The first is to live with it. In that case it would be neither true nor false that the cat at ten minutes past two is intact. This would be to have a somewhat deflationary attitude towards material objects, namely a rejection of bivalence (see chapter 1). This is an option we could defend by pointing out that neither inference nor non-inferential observation, but only prejudice would make us *insist* upon bivalence. But it is still disturbing.

The second option would be an escape from the difficulty by means of a suitable interpretation of probabilities. If we take the 'probabilities' as merely the approximate frequencies which would occur if we repeated a situation often enough, then we need have no fear of the cat. In a large number of similar acts by the suburban guerrillas of A.R.M. approximately 50 per cent of cats would be exploded, 50 per cent left intact. On this interpretation the state of the system does not seal the fate of individual cats. Any particular plaster pet is either exploded or intact prior to observation.

The third option is to insist that I have confused the domains of Schroedinger and von Neumann dynamics. It might be suggested that a good approximation to von Neumann dynamics holds wherever we have significant interaction with macroscopic systems such as plaster cats, even in the absence of observation. I reject this third response for two reasons. First, the macroscopic character of the cat merely serves to *dramatize* a puzzle which occurs already at the sub-microscopic level. There should not be one law for the little and one for the great. It is just as peculiar if very small objects, such as a molecule of calcium sulphate, suffer from the failure of bivalence which Schroedinger's Cat threatens. The macroscopic character only becomes relevant if the theory threatens us with a situation in which we would be able to *see* that the cat is in a limbo of no determinate position. For that would be even more worrying than the case where no such observation is possible. But there is no threat of observing a lack of determinate position: the theory tells us that observation results in determinacy.

My second reason concerns the lack of a sharp division between the macroscopic and the non-macroscopic. So even if the appropriate dynamics for interaction with the macroscopic is not the Schroedinger dynamics, it cannot be exactly von Neumann dynamics either. We must expect a continuum of intermediate cases in which increasing size and complexity lead to more and more von Neumannish dynamics. (I shall comment on this possibility in the next section.) Hence the fate of the plaster cat would not become determinate straight away, nor would it ever become completely determinate. A fraction of a millisecond after the intended time of the explosion the cat will be in limbo. And an hour later there will still be a very small probability that it will be exploded even if it is tending towards the intact state. And this raises precisely the same puzzles as before, only, once again, not so dramatically. This third option then merely removes the drama, it does not affect the metaphysics.

A fourth option, which will be taken up in chapter 7, is that when we observe the cat we determine not just whether it is now intact or exploded but whether it has been since two o'clock. I confess to a fondness for backward causation (Forrest, 1985b).[3] No doubt the reader will react to this with horror.

II THE PROBLEM OF THE TWO DYNAMICS

As usually formulated, quantum theory involves two quite different sorts of dynamics. Sometimes systems evolve in accordance with the Schroedinger dynamics. At others an observation is made, in accordance with von Neumann dynamics.

This raises three difficulties. The first and most serious is a threat of *incoherence*. The second, is the problem of *demarcation*. The third, and, I think least serious, is the *complexity* of having two dynamics for two different situations: our intuitions are that there should be a single unified account of the physical world. I shall discuss these three problems in turn.

First the threat of incoherence. There is no inconsistency in the *formalism* because the different dynamics concern different situations. But the observations being considered are not made with a naked eye, let alone a naked mind. They are made using a complicated apparatus which must interact with the system being observed. And the apparatus is not made of some special stuff. It obeys whatever laws other complicated macroscopic systems obey. As Putnam puts it: measurements are a subclass of physical interactions – no more or less than that (Putnam, 1965, p. 77). As such they should obey Schroedinger dynamics: and here the threat arises. For observations are said to obey the apparently quite different von Neumann dynamics. Admittedly such observations could be *idealizations*, but we have the threat of incoherence between the following prima facie plausible statements:

(1) The complex systems involved in making observations evolve in accordance with Schroedinger dynamics.
(2) For no system, however complicated, does the Schroedinger dynamics provide even a good approximation to the von Neumann dynamics.

There are two different versions of this threat of incoherence. For, as I remarked in chapter 2, the von Neumann dynamics implies both the Strong and the Weak Projection Postulate. It is convenient to consider the threat of incoherence posed by the Weak Projection Postulate first. I shall call this the Threat of Lack of Statistical Coherence. So we shall be considering a large collection of similar systems.

To examine this threat of incoherence in greater detail I shall follow von Neumann (1955, ch. VI) and consider three systems. System II is the apparatus used to measure some quantity of system I, say whether a particle is in region V. System III contains the observer herself. We shall also be considering the composite systems $\mathrm{I} \times \mathrm{II}$ consisting of system plus apparatus, and $\mathrm{I} \times \mathrm{II} \times \mathrm{III}$ consisting of system plus apparatus plus observer. Suppose we are concerned with what happens when I is in a pure state s_0, and a certain observation is made using II. Then the state of $\mathrm{I} \times \mathrm{II}$ will be a superposition of states of the form $s_0 \times u$ for various possible pure states u of II. But the superposition of states of that form is itself of the form $s_0 \times u$.

So we may take the initial state of I × II to be $s_o \times u_o$. The state u_o is one of the possible states of II for which the apparatus is ready for use. In particular I shall assume there is a needle on a dial, a digital display, or some other device which gives the result of the measurement. And I take this to be in the zero position. We now consider a collection of copies of I × II each in the state $s_o \times u_o$.[4]

Provided the composite system I × II is isolated from outside interference, a pure state would evolve, according to the Schroedinger dynamics, into a *pure state* of the composite system. But, according to the von Neumann dynamics, the collection should satisfy the Weak Projection Postulate, and so evolve into a *mixture* of states $s_1 \times u_1$, $s_2 \times u_2$ etc., where each of the u_i is an (approximate) eigenstate for the YNO of whether the position of the needle is the *i*th position.

In some cases, at least, the mixture and the pure state are not even approximately indistinguishable.[5] Does this prove that quantum theory lacks statistical coherence? No. We supposed that the system I × II was isolated, but it is not. For the naked eye observation can affect the system I × II. In particular it may well result in the required transition from a pure state of I × II to a mixture.[6] Von Neumann (1955, ch. 6) assumed that the naked eye observation of the apparatus would itself be an idealized measurement of quantum theory. Given that assumption, he could compare the result of the von Neumann dynamics for I when subject to interaction with II × III, on the one hand, with the Schroedinger dynamics for I × II followed by the von Neumann dynamics for I × II when subject to interaction with III, on the other. He argued that for a suitable interaction between I and II we obtain exact agreement. This seemed to show that quantum theory was statistically coherent after all. However Wigner (1952), and Araki and Yanase (1960) have argued that von Neumann's postulated interaction is not physically feasible. If so a *threat* of incoherence remains, even though there is no *proof* of incoherence.

Some responses to this problem, such as that of Daneri, Loinger and Prosperi (Prosperi, 1971), assume, as far as I can tell, that there is an extra factor not described by orthodox quantum theory. This factor would ensure, or just be, a superselection rule excluding the superposition of various macroscopically distinguishable states. One difficulty here would seem to be that merely excluding the superpositions that should result according to the Schroedinger dynamics does not ensure that the state evolves into a mixture of permitted states. Something else is required to ensure this. Perhaps this something else can also be postulated. If we fiddle enough with the theory we shall obtain the required results. But all this is ad hoc and so to be avoided if possible. And it can be avoided. For von Neumann set himself an excessively tough problem. He assumed that we required exact agreement rather than merely approximate agreement between the naked eye observation of I × II, on the one hand, and the observation of I using II, on the other. The results of Araki and Yanase (1960) and Wigner (1952) strongly suggest that approximate agreement is possible provided the apparatus is large enough. And, of course, every measuring apparatus is macroscopic and so large enough. So, provided these somewhat provisional results are not overthrown by further analyses, the Threat of Lack of Statistical Coherence has been met. To sum up, while the von Neumann and Schroedinger dynamics differ in their effects on the collection of composite systems I × II, the further effect of the observer herself obliterates the differences to a good approximation, thus restoring the statistical coherence.

There remains the threat of incoherence due to the Strong Projection Postulate, which I call the Threat of Lack of Individual Coherence. Let us assume that the Schroedinger dynamics results in a good approximation to the von Neumann dynamics when we consider the effect on a large collection of systems. It remains the case that when a single system is observed, *it* evolves in a probabilistic fashion, quite unlike the deterministic evolution of the Schroedinger dynamics. It is convenient to consider this Threat of the Lack of Individual Coherence, at the same time as the two other problems: the Demarcation Problem and the Complexity Problem.

We shall consider five different proposals listed in order of increasing radicalness:

(1) The Schroedinger and von Neumann dynamics hold in two precisely defined domains.
(2) Von Neumann dynamics, or an approximation to it, may be derived from Schroedinger dynamics (and other assumptions).
(3) Both von Neumann and Schroedinger dynamics, or approximations to them, may be derived from a single underlying dynamics.
(4) Von Neumann dynamics never holds, but a suitable substitute for it does.
(5) Von Neumann dynamics never holds, and we do not need a substitute.

First, then, the two dynamics could be autonomous, but concerned with different domains. This would not only enable us to avoid the above, somewhat inconclusive argument for statistical coherence, but it would also restore individual coherence. However, it would inevitably be more complex than some of the other proposals. One suggestion is that Schroedinger dynamics holds unless there is interaction with the macroscopic. For sufficiently large systems something like classical mechanics is to hold; for sufficiently small ones, Schroedinger dynamics holds. And, where there is interaction between the larger and the smaller, von Neumann dynamics holds. In that case it would be inappropriate to treat system II, the apparatus, in a quantum-theoretic fashion, namely as undergoing Schroedinger evolution. But there is an obvious, and overwhelming objection to this proposal – if it is taken as a serious piece of metaphysics, which it probably seldom is. The macro/micro distinction is one of degree. But if we have two, and only two, autonomous dynamics, then we shall have a precise demarcation, which is not just a matter of degree.

More promising is the dualist version of the first proposal, a version defended by Wigner (1962) among others. According to Wigner, Schroedinger dynamics governs the physical world provided there is no interaction between it and the mental, which in this case is taken to be non-physical. Since we may distinguish the mental from the physical, we have solved the Demarcation Problem.

The obvious difficulty with this proposal is that it involves a special complex psycho-physical interaction law governing the effect of the mental on the physical in observation. The need for such laws is widely regarded as a problem for dualists. It might seem an objection, then, to this proposal for handling the Problem of the Two Dynamics that we are stuck with psycho-physical interaction laws which we would rather avoid.

The second strategy is to derive (an approximation to) von Neumann dynamics from the Schroedinger dynamics. We can do this if we give a statistical interpretation for even the pure states. Consider a one-particle system. On this speculation a quantum state describes a large collection of one-particle systems, rather than a single system. And, among other things it describes the proportion of members of the collection with particles in region *V*. This quantum state evolves in accordance with Schroedinger dynamics. And if we considered a large collection of systems each with an observing apparatus and an observer, the observation process would still be governed by the Schroedinger dynamics.

Assuming statistical coherence, it follows that the large collection of systems will evolve, under observation, approximately in accordance with the von Neumann dynamics. (That is, the Weak Projection Postulate is satisfied.) But an actual observation is, we may assume, made by an observer in just one member of the collection. And the frequency of a given result will approximate the probability given by the von Neumann dynamics. Hence the Strong Projection Postulate follows from the Weak if we give a statistical interpretation of all quantum states. This removes the Threat of Individual Incoherence. We have also solved both the Demarcation and Complexity Problems. For there is only one autonomous dynamics.

The third proposal has been advocated by Cartwright (1983), who suggests a more general theory of which Schroedinger and von Neumann dynamics are (approximate) special cases.[7] Or again, we could take Daneri, Loinger and Prosperi to be proposing some extra factor which is negligible in typical microscopic cases, but becomes significant in typical macroscopic cases. This extra factor will tend to prevent superposition of macroscopically distinguishable states, or more simply, prevent states of highly indeterminate location.

I reject this as premature. Not until we have exhausted all solutions based on quantum theory as currently formulated should we adopt such expedients. No doubt quantum theory is not the last word. But the next word, I suspect, will be even stranger than quantum theory. The history of physics provides no comfort for conceptual conservatives.

The fourth and fifth proposals are to deny that von Neumann dynamics ever occurs, but to explain why it *seems* to. These two are modifications of quantum theory, but they are not additions so much as subtractions, and so much more acceptable. If tenable they clearly solve all the problems. I shall now discuss the two proposals in turn.

I assume that the brain is governed by the same laws as the rest of the material world. And I trace the process of observation back into the brain. We already have the observed system I, the apparatus II and the observer III. I now divide III into a part responsible for perception, III_p, and a part responsible for introspection, III_i. This is highly idealized but will serve. I shall ignore introspection of introspections: they would complicate but not significantly alter the account.

Suppose we are to find out whether a particle is in region *V*. There will be eigenstates for systems I, II, III_p and III_i for the YNOs of whether the particle is in *V*, whether the apparatus II says it is, whether III_p records that II says it is and whether III_i records that III_p records that II says it is. I shall suppose that these eigenstates for III_p and III_i are either correlated or identified with mental states of perception and introspection. Common sense tells us that the actual mental state is some such state. But the Schroedinger dynamics will lead instead to a superposition of cigenstates for III_p and for III_i.

On the fourth proposal the physical world is, even when observed, governed by the determinate Schroedinger equation. The von Neumann dynamics is an illusion due to the way that the mental states are correlated with the possible physical states. For a given brain state there will be a probability of a given mental state. But on this proposal there will be no corresponding change of brain state. This proposal stands to the dualist version of the first proposal as Epiphenomenonalism stands to Interactionism. The effect of the physical on the mental is the same, but the mental no longer affects the physical.

I cannot see that this has any advantages to recommend it over the dualist proposal discussed above. And there will be difficulty in ensuring that the mental state produced on one occasion by the physical is appropriate to that produced on the next occasion. To bring this out suppose an observation is repeated. And suppose that the physical state has not been altered the first time. Since the interaction is probabilistic we have no reason to expect the same mental state to occur. For this reason I dismiss this proposal. Its only function is to prepare the way for the fifth.

The fifth, most radical, proposal is available to materialists as well as dualists. We can insist that the mental is always either correlated or identified with the physical, but deny the common sense view that the mental state is one in which the person perceives the particle is in *V* or perceives it is not. We can push the indeterminacy right through the mental. This piece of heroic metaphysics is mind-boggling – but *that* is no objection. What *is* a prima facie objection is that I know on introspection that I seem to perceive that the particle is in *V* (or that the apparatus says it is). But we may now ask: what 'self' is it which knows this? The knowledge of what I am doing derives from introspection. So what is known to be the case is encompassed by one act of introspection. What is not so introspected is not known, but might still occur. Hence as far as the evidence of introspection goes, there could be a different 'self' for each eigenstate of III_i. Each such eigenstate would correspond to what is ordinarily taken to be the whole of the introspection associated with the brain in question. This does not prevent the other eigenstates corresponding to introspections of other 'selves'. In a superposition of eigenstates of III_i the brain state would be correlated with what could only be described as a superposition of different 'selves'. Each one of them would indeed introspect just what common sense tells us it does, namely that a determinate perception has been made.

On this proposal the von Neumann dynamics (i.e. the Strong Projection Postulate) would be illusory. The probability of the state going into a given eigenstate according to von Neumann dynamics would be interpreted as the relative strength in the superposition of the eigenstate corresponding to a correct but partial introspection that one has perceived the apparatus give the result in question. This heroic piece of metaphysics cannot be rejected by appeal to introspection. For it does not deny the results of introspection. It merely denies the claim that introspection can reveal the whole truth about the mental life associated with a given brain.

This proposal is similar to that of Everett (1957), if his views are suitably interpreted. Everett solved some of the problems posed by quantum theory by postulating *many worlds*. This could be a version of the statistical speculation, and de Witt (Everett et al., 1973, pp. 115–65, pp. 167–218) seems to have interpreted it this way. In that case each state would describe a collection of systems in different worlds. I discuss this in chapter 5. However, Everett's original work can also be interpreted as saying that there are many partially correct views of the world rather than many worlds. On this interpretation, he is supporting the fifth proposal. For each partially correct view would be that of one of the many 'selves'.

Of the responses to the Problem of the Two Dynamics I have dismissed the fourth; I would rather avoid the third, with its ad hoc flavour; the first will be taken up again in chapter 8. Assuming we can meet the Threat of Statistical Incoherence, there is an especially neat proposal (the second) which is only available on the statistical interpretation of even pure states. The only solutions available on the Wave Speculation will be seen to be the dualist account and the extremely radical fifth proposal. This is also the only materialist account available if we reject the statistical interpretation. But, in spite of its radicalness, it is, I think, tenable. So there is always *some* tenable solution.

III VERIDICALITY

I say that an observation is *veridical* if the result is true of the system immediately prior to the observation. For instance, consider the observation of the position of a particle. It is veridical if, just before the observation, the particle has the position indicated by the measuring apparatus.

If observations are always veridical regardless of the state of the particle, then I say that Strong Veridicality is satisfied. It is common sense – and, no doubt, a presupposition of science – that naked eye observations are (for the most part and approximately) veridical. Strong Veridicality was thus a reasonable idealization to make earlier in the history of physics. But we should not automatically carry it over to quantum theory. Rather we should think of veridicality as something to be grateful for when it occurs. Therefore a case has to be made before Strong Veridicality is accepted. I can think of four arguments that might be made for it. While I doubt whether all these have been seriously proposed, it helps to discuss them in order to show that there is nothing peculiar about the rejection of Strong Veridicality.

The poorest argument is that, by definition, observation is an activity for which Strong Veridicality holds. As always we can reply to such attempts at linguistic repression by allowing that the word 'observation' might be used that way. However, I shall use it in a different way. An activity which gives us answers to YNOs which may only be right some of the time will still be called *observation* provided we have no better guide.

Almost as poor an argument is the claim that part of the realist tradition I wish to uphold is the claim that Strong Veridicality holds. Maybe I am not a 'true realist'. So what? I have made it clear enough what I mean by *realism* and what I am defending.

A more serious argument could, perhaps, be based on Instrumentalism. It could be said that there is nothing more to a particle's having a given location than its being observed to have it. Hence, Strong Veridicality must hold. This would be misleading. For in the context of a probabilistic theory of observations it seems better for the Instrumentalist to say that there is no position prior to observation and hence that Strong Veridicality fails.

Anyway, the Instrumentalist begs the question. To be sure, if we fail to provide a 'realist interpretation' of quantum theory, then *that* might be an argument for Instrumentalism. But no one should assume Instrumentalism until then. For it is not a prima facie plausible position (see chapter 1).

The fourth argument requires the premiss that in the classical limit observations must be, to a good approximation, veridical. Even if we grant the premiss, the conclusion does not follow. For provided the deviations from veridicality become negligible as we approach the classical limit, we do not require Strong Veridicality. For any given system we can find a smallest positive number δ such that the probability of an error greater than proportion δ (i.e. $100 \times \delta\%$) is itself less than δ. It suffices that in those systems for which classical mechanics is a good approximation this number δ is negligible. Yet Strong Veridicality would require us to have δ equal to zero for all systems.

I conclude that no case can be made for Strong Veridicality. However, there must be *some* sort of veridicality constraint. My argument for this is that without any veridicality constraint at all there would be no justification for considering a given observation to be one of, say, position rather than, say, energy. Indeed without some veridicality requirement we have no more right to label an observation one of the position of an electron than to label it one of whether there is a ghost next door.

Although this is a simple and obvious argument, it is worth considering objections to it. First, it could be suggested that 'observation of position' (or 'observation of momentum') is a misdescription of what is going on, because we are really observing something subtly different which approximates position (or momentum) in the macroscopic limit. I leave this open as a possibility. Indeed for the case of dispositional momentum I sometimes appeal to misdescription theses. If a misdescription thesis is held, then we should retain a veridicality requirement, but restate it in terms of what really is going on.

A more common reaction would be that of idealists and phenomenalists. A phenomenalist claims that even the word 'observation' is misleading. Perhaps so-called observation is simply a matter of having the appropriate sensations or having a tendency to have the appropriate beliefs. So if I 'observe' a particle's position as near q that would merely be a matter of having a tendency to believe there is a particle near q, or of having the sensations appropriate for a particle near q. Observation would then be denied in favour of belief or sensation. Physicists often say this sort of thing. But it is sad stuff. First, consider the tendency-to-believe account. In something so *recherché* as the theory of subatomic particles, our beliefs do not occur independently of a theory. We only have them as a result of accepting, among other things, quantum theory itself. And to say that what the theory describes is simply the beliefs we tend to have if we accept the theory is a convoluted absurdity.

Nor is the story in terms of sensations any better. Suppose we did treat observations as the having of various sensations. Then quantum theory would become a Kantian theory about states of which we know nothing except that they produce various sensations.[8] Now quantum theory is primarily a theory of the general, or universal. It does not provide predictions that we make particular observations, but rather that we make some observations of a specified *type*, such as an observation that the particle is at q. Indeed you and I could not make the *very same* observation, precisely because we are different observers; but quantum theory is not concerned with such biographical details. So the Kantian theory would have to be one about various *types* of sensation. And there lies its downfall. I ask: In what way are we to classify sensations into types? First suppose there were no restrictions on classifications. Then we could introduce a spurious order into *any* body of phenomena. Various YNOs are made at various times. Sometimes the answer is Yes, sometimes No. If we allow a time-dependent classification then we could say that all the answers were of type T, where type T is defined as Yes at this time, No at that time, as required to fit the data. Thus we would have the enviably simple theory that the answer to any YNO is one of type T.

Obviously we should exclude such gerrymandered classifications, reminiscent of the philosophers' terms of art 'grue' and 'bleen'.[9] But it is not just a matter of excluding classifications dependent on space and time. For, almost certainly, there will be some other features of observations which distinguish them. For the sake of definiteness, suppose every piece of apparatus is slightly different and, because of wear, slightly different on each occasion of use. Let the observations be labelled as number one, number two etc. And let the distinguishing feature of observation number n be that it was performed by an apparatus with feature F_n. Then we can consider the type T^* of observation which is Yes when the apparatus has F_n for some integers n and No when it has F_n for other integers n, where the integers are selected so that once again we have the theory that all observations are of type T^*.

The moral of all this is that when evaluating a theory it is not enough that it look simple in terms of the types used to state it. I require also that the types not be gerrymandered.[10] Now quantum theory is only tolerably simple if the data are classified into types such as *making an observation that the particle is in region V*. Such a classification into types is not available to the proponent of the Kantian speculation, who should classify observations only in terms of the sensations produced. And given *that* way of classifying observations into types, the observation that a particle is in V would turn out to be a highly complex gerrymandered type.

The Kantian theory, then, is no way to avoid a veridicality requirement. But, in desperation, one might propose something even more radical. Perhaps we could deny that so-called quantum 'measurements' are anything more than a kind of process which we foolishly took to be measurements or observations of some kind. Suppose we repeatedly 'measure' the location of a particle or 'particle'. Then we obtain with various frequencies various 'results' of 'measurement'. And on this radical proposal, what we have observed is merely a certain process one of whose products is the generation of these 'results'.

Zen masters, we are told, sometimes unexpectedly beat their disciples while they are deep in meditation, thus drawing them back to the more mundane world. There are times when we would like to do violence to certain thinkers to bring them back to their senses. This is one of those times. A measuring instrument, say a cloud chamber, is not a black-box whose structure is hidden from us. It is a device constructed for a purpose, using a certain amount of theory. The cloud chamber, for instance, is made to detect whether a charged particle was present. The only plausible explanation of why it behaves as it does is that something sufficiently particle-like to have a fairly determinate location was indeed present. And we cannot avail ourselves of that explanation without conceding that the cloud chamber is a way of measuring location and not a way of 'measuring' 'location'.

We need some sort of veridicality requirement then. The weakest I can think of is that for suitable[11] YNOs *Weak Veridicality* is satisfied.

Weak Veridicality, called the *Eigenvector Rule* by Redhead (1987, p. 73), may be expressed thus:

> An observation of X is said to be *weakly veridical* if, when the system is in an eigenstate for X, the recorded result is true of the system just before the observation.

An immediate consequence of Weak Veridicality combined with quantum measurement theory is that, whatever the state *prior to* the observation, the recorded result is true of the system as it is *after* the observation. And such *post facto* veridicality is equivalent to Weak Veridicality. For an eigenstate for a given observation is not affected by that observation.

The converse of Weak Veridicality is worth commenting on.[12] Consider an observation. It satisfies the converse provided:

> The recorded result is true of the system just before observation only if the system is an eigenstate for the observation.

This would follow from Strong Veridicality, and like that principle, would seem to have no a priori justification. To decide whether it holds we should first decide which metaphysical speculation is correct and then see what follows. This is something I shall discuss at the appropriate places in chapters 5, 6 and 7.

There are two other veridicality theses of interest. One of these I call *Mean Veridicality*. It is appropriate only where the quantity to be measured has a determinate value. Let us consider the results of observing this quantity. There will be a probability distribution for such results. Then Mean Veridicality is satisfied if the mean value for that probability distribution equals the actual value of what is being observed.

If there is no determinate value which is being observed we might, instead, have a *probability distribution* for the values of the observable *prior* to observation. We shall also have a probability distribution for the *results* of observation. I say that Statistical Veridicality holds if these two distributions are the same. I call this *Statistical* Veridicality because the case I have most in mind is one in which the probability distribution prior to observation is interpreted in terms of frequencies. Then Statistical Veridicality implies that (very probably) the frequency distribution for the results of observation approximates that for the quantity being observed.

Mean Veridicality or Statistical Veridicality, if appropriate, could hold even though Strong Veridicality fails. For observation may well disturb the system being observed, resulting in non-negligible errors, thus preventing Strong Veridicality. Yet these errors may well 'cancel out' in the sense that an error $+\delta$ has the same probability as one of $-\delta$, for any δ, resulting in either Mean or Statistical Veridicality.

However, if there is a systematic bias in the errors due to the disturbance of the system we should expect Statistical Veridicality or Mean Veridicality, whichever is appropriate, to fail. Let me emphasize this. It is not merely that we have no guarantee that it holds. Rather we should *expect* it to fail.

Consider, for example, an observation of kinetic energy. This is never negative. And that introduces a bias in the disturbance due to observation. While the bias occurs quite generally it is exhibited most clearly by cases in which the actual kinetic energy is zero. In that case the observed kinetic energy could well be $+\delta$, for some positive δ, but never $-\delta$.

A similar bias is to be expected for spin. Recall that the only spin observations are of whether the spin is 'up' or 'down' relative to a given direction as 'straight up'. One interpretation of this is that the particle has an actual spin direction making a specified angle, α, from $0°$ to $180°$ with the straight-up direction, and that what we are observing is whether this angle is acute or obtuse. We should expect a bias in the disturbance to the angle. Once again this bias is quite general but is most clearly exhibited in the extreme cases, where $\alpha = 0°$ or $\alpha = 180°$. In those cases there can only be errors one way.

Mean or Statistical Veridicality will not, then, be even a desideratum except for those observations such as position or momentum for which we should expect the errors to cancel. This result is an important one, of relevance to the discussion of so-called No Hidden Variables Theorems. Hence it is worth considering an objection to it. This is that we might be able to stipulate, say, Mean Veridicality for kinetic energy by adjusting the observed results, so that some were negative.

In reply I note that this stipulation is in several ways incompatible with the usual formalism (which is assumed when proving the No Hidden Variables Theorems). We would not, for instance, have repeatibility for observations. Nor could the results of observations for the kinetic energy be obtained by the standard formula $(p^2/2m)$ from those for momentum, p. For if they were, then the results would never be negative.

IV THE TWIN-SLIT EXPERIMENT[13]

We are to imagine a screen with two slits which can be shut if necessary. A beam of electrons is aimed at the screen, behind which there is an electron detector (say a photographic plate) (see figure 3.1).

One notable phenomenon is that the distribution of electrons on the detector exhibits an interference pattern reminiscent of that obtained by Young in his classic experiment when he passed light through a screen with two pinholes (see Born and Wolf, 1980, pp. 260–1) (figure 3.2). The frequency, like the intensity of the light

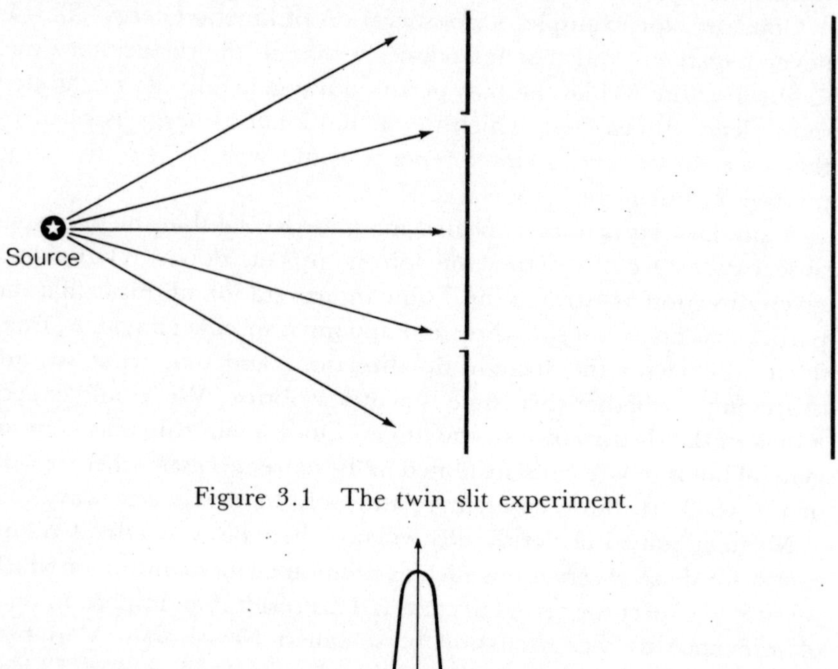

Figure 3.1 The twin slit experiment.

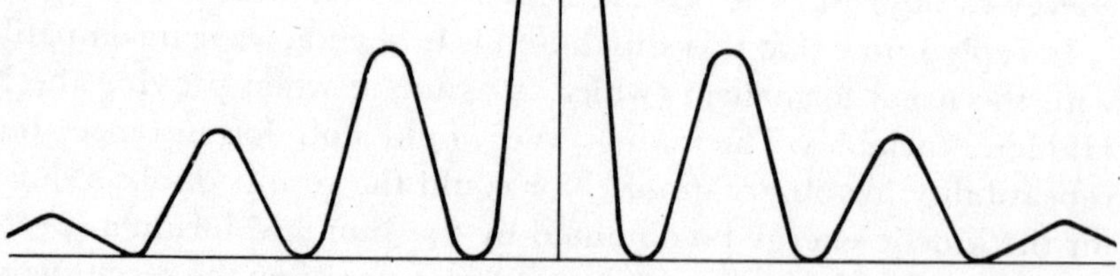

Figure 3.2 Results of the twin slit experiment.

in Young's experiment, has crests and troughs, the troughs specifying regions where electrons seldom reach. This is highly suggestive that something wave-like is going on, the crests being where the waves from the two slits reinforce, the troughs where they cancel out. Moreover if we arrange the slits suitably we can ensure that there is a trough precisely where, if one slit were shut, there would be a peak.

The twin-slit experiment is puzzling, then. Any speculation about quantum theory should explain what is going on. Not surprisingly the peculiarities of the twin-slit experiment involve the superposition of states. Consider the YNO of whether the electron went through slit one, and the YNO of whether it went through slit two. Suppose v_1 and v_2 are eigenstates for these YNOs respectively. Then the state of an electron in the twin-slit experiment will be a superposition of v_1 and v_2. And while we can interpret v_1 and v_2 as states in which the electron goes through slits number one and number two respectively, our puzzlement derives from our lack of any obvious interpretation of the intermediate cases where we have superpositions. Where superposition is otherwise puzzling we can always reach for a wave interpretation. For the superposition of waves is not puzzling at all: it is merely the result of taking suitable combinations of the field strengths.

I have described the twin-slit experiment merely as a puzzle. But it is often considered to be not simply a puzzle but rather a strong argument against realism (in my sense) about electrons considered as particles. To state this argument we need some notation. The argument involves probabilities. And since we shall be concerned both with *events*, such as a given particle's going through the first slit, and quantum *states*, it is convenient to consider probabilities of *propositions* relative to other *propositions*. For then we can handle both events and states. Let 'Pr(h/k)' denote the probability of h given k *and* given some statement describing both quantum theory and the twin-slit set-up.

We consider a given electron e passing through the twin-slit set-up. For any quantum state u of that electron we can consider a proposition Eu, namely that this state *obtains* or *exists*. Let S_1e, and S_2e, be the propositions that the particle e passes through slits one and two respectively. Let Re be the proposition that the electron arrives at some designated region R in the screen.

Then the assumption that the electron is a particle implies:

(i) $\Pr(S_1e \text{ or } S_2e/Eu) = 1$, for any state u.

and:

(ii) $\Pr(S_1e \text{ and } S_2e/Eu) = 0$.

I shall assume distributivity, namely:

(iii) [*Re* and (S_1e or S_2e)] is equivalent to [(*Re* and S_1e) or (*Re* and S_2e)].

Putnam (1965, p. 181) has sought to avoid the purported reductio by denying (iii). But I claim that this is unnecessary. For (i), (ii) and (iii) are, I believe, jointly consistent with quantum theory.

Now suppose v_1 and v_2 are eigenstates for the YNOs of whether the electron went through slit number one or slit number two respectively. And suppose u is a superposition of v_1 and v_2 making the same 'angle', namely 45°, with v_1 and v_2. Then:

(iv) $\Pr(S_1e/Eu) = 1/2 = \Pr(S_2e/Eu)$.

From (i), by the standard Calculus of Probabilities we have:

(v) $\Pr(Re/Eu) = \Pr[Re \text{ and } (S_1e \text{ or } S_2e)/Eu)$.

From (iii) and (v), we have:

(vi) $\Pr(Re/Eu) = \Pr[(Re \text{ and } S_1e) \text{ or } (Re \text{ and } S_2e)/Eu]$.

From (ii) and (vi), we have:

(vii) $\Pr(Re/Eu) = \Pr(Re \text{ and } S_1e/Eu) + \Pr(Re \text{ and } S_2e/\mathrm{Eu})$.

From (iv), using the Multiplication Rule for Probabilities, we have:

(viii) $\Pr(Re \text{ and } S_ie/Eu) = ½\Pr(Re/S_ie \text{ and } Eu)$, for $i = 1,2$.

From (vii) and (viii), we have:

(ix) $\Pr(Re/Eu) = 1/2\Pr(Re/S_1e \text{ and } Eu) + 1/2\Pr(Re/S_2e \text{ and } Eu)$.

In fact, quantum theory (Audi, 1973, p. 130) tells us that:

(x) $\Pr(Re/Eu) \neq 1/2 \Pr(Re/Ev_1) + 1/2 \Pr(Re/Ev_2)$.

By Weak Veridicality, Ev_1 entails S_1e. So we may re-write (x) as:

(xi) $\Pr(Re/Eu) \neq 1/2 \Pr(Re/S_1e \text{ and } Ev_1) + 1/2 \Pr(Re/S_2e \text{ and } Ev_2)$.

(ix) and (xi) would lead to a contradiction if, for i = 1 and 2, (S_ie and Eu) were equivalent (given the background information) to (S_ie and Ev_i).

Now u is a superposition of v_1 and v_2. The objector, I suspect, has assumed that *being in state u* is equivalent to the disjunctive state: *being in v_1 or being in v_2*. In that case (S_ie and Eu) would be equivalent to [S_ie and (Ev_1 or Ev_2)]. But, by Weak Veridicality, S_1e is inconsistent with Ev_2, and S_2e is inconsistent with Ev_1. So we would then have the required equivalence and hence the contradiction.

It is worth noting that in Putnam's discussion (1965, pp. 177–81) the quantum logic is the logic, or 'logic' of the YNOs. And this has the advantage that there is a unique disjunction. However, in the problem of the twin-slits we need to consider *states* not YNOs. Furthermore, we are considering a state which is a *superposition* of v_1 and v_2 (Audi, 1973, pp. 129–30). And there is a good reason why we cannot identify such a superposition with a disjunction, whether classical or of the so-called quantum logic. It is that there is not even a *unique* superposition of v_1 and v_2 as required, if the superposition is to be interpreted as disjunction. The proposed equivalence of *being in u* with the disjunctive state *either being in v_1 or being in v_2* is, I suspect, the result of confusion. So Putnam has conceded what he need never have conceded, namely that with distributivity we would have an objection to taking electrons as particles. For that reason, no further discussion is required of distributivity. For we have already seen in chapter 2 that the non-Boolean character of the so-called quantum logic provides no reason for denying distributivity for conjunction or disjunction.

V POSITION AND MOMENTUM AGAIN

In chapter 2, I mentioned two peculiarities concerning position and momentum. The first was Heisenberg's Uncertainty Principle.

The second was the difficulty of getting positive joint probabilities for position and momentum. But the most important question has yet to be tackled. Can a given electron have both a precise (i.e. determinate) momentum and a precise position at a given time? Relevant to this is whether it can be shown that there is a way of veridically measuring both position and momentum together. For, if there is, then electrons must sometimes have both a precise position and a precise momentum. It would then be ad hoc to deny that they always do. If there are no such veridical measurements then we are free to *speculate* that an electron does not have both precise position and precise momentum.[14] Recall that I have not assumed Strong Veridicality. The only direct constraint on the veridicality of standard measurements is Weak Veridicality, namely that in an eigenstate for a YNO that YNO is veridical. This requirement does not establish that position and momentum can be jointly measured in a veridical fashion. For there are no quantum states which are both eigenstates for being a finite region V and for having a momentum in a bounded set W of momenta.[15]

The only other possibility for the veridical observation of both the position and momentum of a given particle at a given time is in some indirect way, not using standard measurements. I shall now examine two thought experiments which might seem to establish that there are such measurements. Both fail. The first is Heisenberg's Concession (1930, p. 20). Heisenberg allowed that if we had already determined the velocity of an electron at one time and later observed its position then we could find the position and momentum of the particle between the two observations, to as great an accuracy as we please. But, for Heisenberg's thought experiment to show what he thought it showed, the measurement of position would have to be veridical. Heisenberg, it appears, was thinking that while measuring position disturbed momentum it would not disturb the position. And that is evidently the situation in classical mechanics, where as Heisenberg argued, we already have interference between the observation of position and momentum. So Heisenberg's argument shows that in classical mechanics we can observe both position and momentum to a given accuracy. But in quantum theory, the measurement of position is not necessarily veridical unless the state is an eigenstate. And it is begging the question to assume anything more than Weak Veridicality. So Heisenberg's thought experiment would require that an electron be in an eigenstate for position just before the second observation.

But we may not assume this. I conclude that Heisenberg's Concession was premature.

Somewhat similar is the well-known Einstein–Podolsky–Rosen thought experiment (Einstein et al., 1935). Here we provide a situation where the positions and momenta of the different particles are correlated. For instance, we can ensure that we know the mid-point between the two positions, and that we know the difference between the momenta. Then the particles move far apart. We measure the position of one and so calculate that of the other. And we measure the momentum of one and so calculate the momentum of the other. In this way we have obtained a result for both the position and the momentum of both particles.

If that was all there was to the thought experiment then it would be open to the objection I made to Heisenberg's concession. We could suggest that measuring the position of one particle affected its position and hence the mid-point of the positions of the two particles. However, we can ensure that the von Neumann dynamics does not affect the mid-point.[16] Hence this objection fails.

The Einstein–Podolsky–Rosen thought experiment fails in a subtler way, however. Consider the YNO A of measuring the position of the first particle and the YNO B of measuring the momentum of the second. To find the position and momentum we would have to make both A and B. But just as the formalism shows that the position and momentum for one particle cannot be measured simultaneously in the standard fashion using YNOs, it shows that we cannot independently measure the position of one and the momentum of the other particle using YNOs, unless, by chance, the state of the two-particle system is decomposable (see chapter 2, section 4).

Here it could be objected that to defend the impossibility of measuring position and momentum simultaneously, we must concede that the measurement of the position of one particle affects the momentum of the other at a distance, and that this is a disadvantage. However, a further discussion (section VII below) of a similar thought experiment shows that we have to reconcile ourselves to something like action-at-a-distance anyway. So this objection fails.

I conclude that we are free to speculate that electrons, and other particles, never have both precise positions and precise momenta. However, this *is* just a speculation. We are also free to speculate that they do have both precise positions and precise momenta. For there are YNOs for position up to any specified accuracy, and these will have eigenstates. So by Weak Veridicality there are electrons with positions of arbitrarily great precision. Likewise there are electrons with arbitrarily precise momenta. And this should blunt any positivist argument that it is *nonsense* to ascribe both position and momentum (within an arbitrary given accuracy) to a given electron. For we cannot reasonably demand the verifiability of the conjunction as well as of the conjuncts. If we do, then such innocuous sentences as 'The match would light if I should strike it, but I shall not strike it' would be nonsense. It is still open to someone to claim that the ascription of both position and momentum to an electron is nonsense on the grounds that it is not the sort of entity which has both. This would be like saying it is nonsense to attribute mass to the square root of two. But that claim would have to be *based* on a speculation in which electrons are said to lack both position and momentum, and so could not be used to *support* such a speculation. Furthermore, in that case I would prefer to say simply that such sentences lack truth-value, rather than say they are nonsense.

I conclude that the question whether particles have both precise position and precise momentum is something we are free to speculate about either way.

VI TWO PROBLEMS CONCERNING MIXTURE

In chapter 2 I noted three possible ways of interpreting *mixture*. The first is the *Statistical Interpretation*. According to this, if the individual systems are in pure states u and v, a mixture $(m/n)u + (1 - m/n)v$ *always* represents a large collection made up of copies of u and copies of v in the proportions m to (n – m). Closely related is the *Ignorance Interpretation*: if an ideally rational thinker who knows all about quantum theory has degree of confidence λ that the state of the system is u and degree of confidence $1 - \lambda$ that it is v, then the appropriate representation for that state of ignorance is $\lambda u + (1 - \lambda)v$. For our present purpose, these two interpretations can be handled together. This is because the Ignorance Interpretation can be assimilated to the Statistical Interpretation by considering large collections of *possible* systems. If necessary, we approximate the degree of confidence λ by a fraction m/n. Then we may (to a good approximation) treat having a degree of confidence λ that the system is in state u as having equal degrees of confidence that the actual system is any one of a large collection of n possible systems of which m are in state u.

The *Sui Generis Interpretation* is quite unlike the other two. It treats mixture as a primitive, unanalysable operation on states. In that case, pure states do not differ in any remarkable way from mixed ones. They are just those states which are not obtainable from other states using 'mixture'. There is still a connection between literal mixture and this 'mixture'. It is that a large collection of states each in a 'mixture' of pure states will be observationally indistinguishable from a literal mixture of systems, some in one pure state, some in another.

In order to discuss which interpretation is appropriate, I shall consider two problems. I call them the Composite Systems Dilemma and the Non-uniqueness Problem.

The Composite Systems Dilemma may be illustrated by considering a two-particle system. It is convenient to consider two different kinds of particle, so we shall consider a hydrogen atom made up of an electron and a proton. Suppose it is in a pure but non-decomposable state which is the superposition of the two decomposable states $u \times v$ and $w \times z$. Then we may ask 'In what state is the electron?' It is not in some pure state. For if it were, then since the atom is itself in a pure state, the atom would be in a decomposable state.[17] But it is not. It is usual, therefore, to infer that the electron is in a mixed state.[18]

The alternative is to deny that the electron is in any state *directly* describable in quantum theory (Hooker, 1972, p. 97). Such a state would be describable *indirectly* as the state of a subsystem of a system in such and such a quantum state. This alternative position may conveniently be referred to as the Irreducibility of Composite Systems.

We may now state the Composite Systems Dilemma: *Either* some states of composite systems are irreducible *or* some subsystems are in mixed states where the composite system is in a pure state. Neither seems particularly palatable. Relevant to this dilemma is the fact, noted in chapter 2, that decomposable pure states can evolve, by Schroedinger evolution, into non-decomposable ones. Hence, either reducible states of composite systems evolve into irreducible ones or pure states of subsystems evolve into mixed states.

The second puzzle is the Non-uniqueness Problem (Beltrametti and Cassinelli, 1981, pp. 9–11; Gibbins, 1983). We can get the same mixed quantum state in more than one way. That is, we can mix states u and v and obtain the same quantum state as if we had mixed quite different states w and z. As Gibbins nicely puts it: 'It is as if one were saying that an urn contained a collection of balls each of which is either black or white, *and* either green or red' (p. 196).

It is convenient to concentrate on a special case of the Non-uniqueness Problem. This illustrates the difficulty without introducing distracting generality. Consider two orthogonal pure states u and v. And consider a pair of orthogonal states w and z each of which is a superposition of u and v. Each of w and z makes an 'angle' of 45° with u and v (see figure 3.3). Then the formalism represents the two mixtures $(1/2)u + (1/2)v$ and $(1/2)w + (1/2)z$ by the same mathematical item.[19]

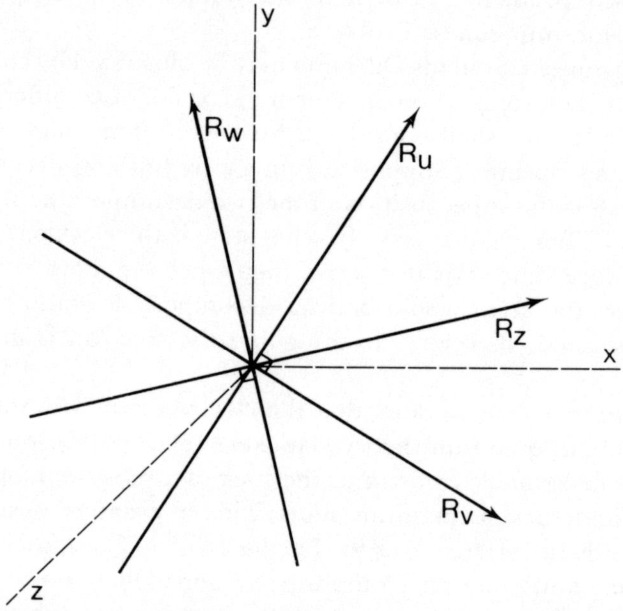

Figure 3.3 The states w and z are superpositions of u and v, and vice versa.

An even more special case concerns spin. Consider the eigenstate of whether the particle is spin-up with respect to a. Likewise consider the eigenstate of whether it is spin-down with respect to a. Then mix these two pure states in equal proportions. The mathematical item representing the mixture is exactly the same no matter which direction a is chosen.[20] A most interesting result.

I shall now use these problems to argue for two theses concerning mixture. The first I call the Uniformity Thesis. It says that if we interpret mixture literally, then the pure states not only *may* but *must* be interpreted statistically, that is as describing large collections of systems not more directly describable in terms of quantum theory. Likewise, it says that if we give an ignorance interpretation of mixture, then we must take pure states as describing various degrees of confidence which an ideally rational person might have in the state of a system not more directly describable in terms of quantum theory. Hence, it implies that we should treat mixture as primitive and unanalysable if we make some other speculation about pure states.

The second thesis is, not surprisingly, that it is an advantage not to have to treat mixture as primitive and unanalysable.

I shall argue for the Uniformity Thesis in two ways. First, I shall suppose we reject the Irreducibility of Composite Systems. The Uniformity Thesis will then follow from an examination of the Composite Systems Dilemma. However, I shall not discuss whether we should or should not reject the Irreducibility of Composite Systems. For this would depend on the further details of various speculations. My second argument is more important, because it does not require the rejection of the Irreducibility of Composite Systems, being based on the Non-uniqueness Problem.

If we reject the Irreducibility of Composite Systems, then there will be a pure state of a hydrogen atom whose electron is in a mixed state. Now, we cannot have a literal mixture of states for the electron without considering a large collection of electrons, and hence a collection of hydrogen atoms. Hence, even the pure state in question not only *may* but *must* be taken to represent a collection of hydrogen atoms rather than a single one. And in case it is objected that this holds only for non-decomposable pure states, it should be remembered that not only do decomposable ones evolve into non-decomposable ones, but, conversely, non-decomposable ones evolve into decomposable ones. The rejection of the irreducibility of Composite Systems leads, then, to the Uniformity Thesis for the Literal Interpretation of mixture. An exactly analogous argument holds for the Ignorance Interpretation.

Now consider the Non-uniqueness Problem. An easy, but not entirely satisfactory solution is to insist that the mixtures considered are different collections which are observationally indistinguishable. Given this solution, we would say that the formalism reflects this indistinguishability. It represents quite different mixtures by the same mathematical item. But no explanation would be available for the indistinguishability. In order to keep mysteries to a minimum, it is preferable to give a solution in which mixing quite different states results in the very same resultant state. One way in which a mixture of *u* and *v* could be the same as a mixture of *w* and *z* would be if mixture was primitive and unanalysable. In that case, there is no reason why different mixtures should not result in the very same state. But there is no reason why they should either. So the problem is only partially solved.

More interesting is the way in which a literal mixture could result in the same state. This can occur only if the pure states themselves describe heterogeneous collections. This sounds paradoxical. But it is not. A pure state is one which cannot be expressed as a mixture of other *quantum* states. That does not prevent *every* quantum state even pure ones from describing a mixture of 'hidden' or non-quantum states. A simple example of how this could solve the Non-uniqueness Problem is the following. Consider four collections of coloured specks. Call them *U*, *V*, *W* and *Z*. *U* consists of 50 per cent red and 50 per cent blue specks. *V* consists of 50 per cent red and 50 per cent yellow specks. *W* consists of 100 per cent red specks. And *Z* consists of 50 per cent blue and 50 per cent yellow specks. These are quite different collections, but if we mix *U* and *V* in equal proportions, we obtain the same result as mixing *W* and *Z* in equal proportions.

Applying this to quantum theory, all we need suppose is that the pure states themselves describe heterogeneous collections of systems. What happens in the case of position and momentum is not straightforward. I discuss it in chapter 5. Here I shall consider only the special case of spin. I take the YNO of whether the spin is up with respect to direction *a* as whether the spin makes an acute angle with *a*. Now consider a pure state which is an eigenstate for that YNO. It may be taken to describe a collection of particles whose spins all make an acute angle with *a*. If we think of each spin as a vector of length 1/2, then these vectors will be (approximately) uniformly distributed around a hemisphere of radius 1/2, whose axis is the direction *a*. Now mix equal proportions of spin-up and spin-down pure states. We obtain an (approximately) uniform distribution around a sphere. And this is now independent of the direction *a*. Problem solved! Does not the explanatory power of this hypothesis provide grounds for accepting it?

A similar argument holds on the Ignorance Interpretation. In both cases the Uniformity Thesis follows by an inference to the best explanation.

The other thesis, that it is an advantage not to have to treat mixture as primitive and unanalysable, is a corollary of the above argument. For if we do treat it as primitive, then we are already committed to the thesis that quite different states of collections of systems may be indistinguishable. (A homogeneous collection of 'mixed' states will be indistinguishable from a heterogeneous one of pure states.) And, as I mentioned above, there would be no explanation for this indistinguishability. Hence, we should prefer hypotheses on which we can avoid it.

VII ACTION-AT-A-DISTANCE

With unaided muscles all we can do is move our bodies, which may push, pull or deform other objects, which may, in turn, affect that which is some distance from us. We cannot directly manipulate what is distant.[21] Is all causation a matter of push and pull? According to the Clockwork Theory of the World, as Popper calls it (1982, p. 163), it could be. But we have no reason to accept that theory.

Yet the merest suspicion that a speculation requires action-at-a-distance has been known to bring about its condemnation. Why is that? I claim that it is because purported examples of action-at-a-distance in quantum theory are also examples of more properly disturbing novelties.

It would be surprising if action-at-a-distance *by itself* was cause for concern. For Newton's theory entailed action-at-a-distance. Thus, on it, I would have the god-like power to affect far-distant stars by merely lifting my little finger. Of course, the effect would not be very large, but, unlike the law, metaphysics cares about very little things. Now Newton and his contemporaries seemed to have worried about this. But we may attribute that to the undoubted simplicity of the Clockwork Theory. If the rival to a theory entailing an action-at-a-distance *were* the Clockwork Theory, then we should prefer the latter. But in the present case it is not. So we should not worry about action-at-a-distance as such.

To illustrate the more properly disturbing novelties associated with action-at-a-distance I consider, in turn, the twin-slit thought experiment, and Bohm's version of the Einstein–Podolsky–Rosen thought experiment (Bohm, 1951).

These problems are starkest if we take electrons, protons etc. to be particles and not mere 'particles'. So I shall make that assumption in this section.

The behaviour of electrons going through one slit would seem to depend on whether or not the other slit is open. With the slit closed a number of electrons could reach a region *R* on the screen which is rarely if ever recorded with the other slit open. Compare this with an experiment in which the other slit is replaced by an electromagnet. If it is switched on it will significantly affect the behaviour of electrons passing through the slit. This is not especially puzzling. For the electromagnet produces a *force* on the electron, and, if desired, this force can be considered to be mediated by an electromagnetic field. In that case there is no direct action-at-a-distance. The puzzle with the twin-slit experiment is that, as an electron goes through one slit there is no significant *force* which is due to the other. Only those forces due to the screen in the immediate neighbourhood of the first slit act significantly on the electron. It is as much the forcelessness as the action-at-a-distance which is distressing. But the two are related. For if the effect were due to a force we might be able to posit a suitable field to mediate the effect, and so remove direct action-at-a-distance.

Is it then a mere prejudice in favour of forces, and in favour of forces acting only on nearby objects, that is threatened by quantum theory? No. There is more than mere prejudice at work. To demonstrate this I shall rationally reconstruct some research programmes into which just about all of science could be retrospectively fitted. My aim is not to do the history of ideas, but merely to explain why scientists value forces and value locality. I begin with the most generic programme, that of *Methodological Atomism*. Its strategy is divide and conquer, or, more accurately, divide and ignore.

To explain any aspects of the physical world in detail, it seems necessary to assume that complex systems can be understood in terms of components and the way these components are related. That the world can be so analysed could be an a priori knowledge claim. But it can more modestly be seen as part of an explanatory hypothesis. And to the extent that this hypothesis explains what is observed and makes successful predictions, it is justified.

Any philosophical thesis that the world is a simple unanalysable entity or, more prosaically, that it is a system whose complexity in practice defies analysis, is a threat to Methodological Atomism. As such a scientist's reaction to it could be predicted to be hostile. This may explain why the scientist, at least in the stereotype, is hostile to mysticism. For what mystics say often lends support to the thesis that 'All is One', which, at very least, implies that the world is a system of which the parts are so intimately related as to defy analysis.

There are two more specific programmes, which I call the Dyadic Programme and the Locality Programme. Both these are quite natural extensions of Methodological Atomism and both are threatened by the twin-slit experiment.

Methodological Atomism is compatible with the complex systems being analysed into simpler ones, where the relations between the simple ones are of arbitrarily high polyadicity. But the analysis will be simpler, and hence the chance of explanation greater, if all the relations are dyadic. The Dyadic Programme is the background assumption to much of modern science. It says that we should analyse complex systems into simpler ones, their properties and dyadic relations between them. The analysis of molecules in terms of atoms and bonds between atoms is one example. Another is Newtonian mechanics in which the behaviour of particles is governed by forces between pairs of particles. Even field theories fit into this programme. The particles affect the field and the field the particles. A Triadic Programme or even a Tetradic Programme might still be simple enough, but if we can avoid going beyond dyadic relations, so much the better.

Another way in which Methodological Atomism can be fostered is by the Locality Programme. This is the assumption that all *non-negligible* immediate effects on one component of the system are due to components which are nearby. What counts as negligible and as nearby depends on the context. The assumption that gravitational forces on planets due to the fixed stars are negligible could be seen as an instance of the Locality Programme even if the force of, say, Jupiter on Mars is not ignored. For in this context Jupiter is *near* Mars.

The twin-slit experiment supports the following destructive theses:

(i) There are significant effects not due to forces.
(ii) The Dyadic Programme must be abandoned.
(iii) The Locality Programme must be abandoned.

In support of (i), the effect of one slit on the particles going through the other is not due to special forces, which are not being posited. In support of (ii) we may note not merely the abandonment of forces but the way in which the whole configuration of the slits affects the particle. Since the phenomena being considered are due to superposition, we may also support (ii) by noting that one state being a superposition of two others is a triadic, not a dyadic relation. In support of (iii) we have the obvious distance between one of the slits and the electrons affected.

First let us consider the love of forces. Part of the resistance to (i) is, no doubt, mere conservatism. But it has been suggested that there might be more to the love of forces than that.[22] One of the most fundamental principles in physics is the conservation of energy. But this can be held only if we have a definite concept of energy. In classical mechanics we have the kinetic energy due to motion and the potential energy due to the forces. The threat of effects not due to forces is that it is not obvious what connection they have to the energy.

This threat can be met by *displaying* the connection between energy and the effects not due to forces. There is, I shall argue, a perfectly clear connection, even though it is not *quite* that of classical mechanics.

First we notice that quantum theory provides an analogue of the principle of energy conservation. Consider the YNO of whether the energy equals E to within ϵ. Then the theory tells us that at each time t there is specified probability $\mu(t)$ that measuring this energy using this YNO gives us a result equal to E to within ϵ. The analogue of energy conservation is that for any choice of E and ϵ, $\mu(t)$ does not change with time, provided the system is undergoing Schroedinger evolution.

In general we would not insist upon energy conservation in the von Neumann dynamics, because in an observation there is outside interference with the system.

Further speculation might suggest stronger analogues of the classical principle.[23] But, *at least* we have the above result. Moreover, the formula for the energy is a close analogue of the classical formula.[24] It shows how the probabilities for energy are determined by the state and the (actual and potential) forces. While not quite classical, we have a clear connection between the forces and that which is conserved.

The solution to the problem of how the non-force effects contribute to the energy is that they do not. The energy is determined by the state and the forces *without reference to any further effects*. However, this solution might be thought of as itself open to the objection that all causation must involve energy transfer, as in the *transference model of causation* (Aronson, 1971; Fair, 1979).

However, the thesis that all causation must involve energy transfer is itself open to a serious objection. It is that even in classical mechanics there is no *proportion* between the magnitude of the energy transfer and the magnitude of the effect. We can find systems for which a given effect can be produced by means of an energy transfer as small as you please (Crutchfield et al., 1986). A simple example is that of a pin-ball machine, with movable pins. The energy required to move a pin is very slight, but it can drastically alter the trajectory of the ball. Because the energy transfer required may be small out of all proportion to the effect, it is probable that the requirement that there be *any* energy transfer at all is a mere accidental characteristic of classical mechanics. If energy-transference were constitutive of causation, we would expect a better correlation between the magnitude of the energy transfer and the magnitude of the effect.

I draw the conclusion that forceless action-at-a-distance in no way threatens energy conservation. It does provide grounds, though, for denying that all causation requires a transference of energy.

Now consider the Dyadic and Locality Programmes. It is not surprising that they should be cherished. For these two *species* of Methodological Atomism might well have been thought of as preconditions of scientific progress. But quantum theory is squarely within the programme of Methodological Atomism, and its immense success could not have occurred if we could not analyse complex systems such as atoms, into simpler ones. So how is such success possible if the Dyadic and Locality Programmes are to be abandoned? Part of the answer is that a few specified triadic relations need not be intolerably complex. For example the relation which the sum has to the summands is a triadic relation. But it is a paradigm of mathematical tractability. And in quantum theory we build up complex systems out of simpler ones using varying novel but perfectly tractable operations such as superposition, all of which could be analysed into triadic relations. In addition we retain the dyadic relation of *force*.

Successful explanation without locality is more perplexing. If even distant items could have significant effects on what is happening nearby how can we guarantee that we shall explain a given phenomenon? The answer here, as elsewhere, is that we must give up the demand for *guarantees*. There is indeed a *threat* of some sort of holistic effect, but if, on a given occasion, we *can* explain what is happening in terms of what is nearby then the success of that explanation, on that occasion, provides grounds for saying that *in this case* violations of locality may be ignored. To illustrate this, suppose that the overall topology of space–time has an immense effect on the way particles behave. Suppose, to make this more precise, that there will be electromagnetic forces only if space has the topology of a sphere. This would be a massive violation of the Locality Programme. But we could still explain the behaviour of particles using electromagnetic forces. And that explanation holds regardless of the immense influence the shape of space is having. Or again suppose the total number of electrons in the universe had to be constant. Then no electron could be annihilated here without one being created somewhere or other. Perhaps that would result in the otherwise inexplicable appearance of electrons from time to time. But we could explain phenomena ignoring this highly non-local effect, and most of the time we would get away with it. The moral is that in order to explain something we do not have to consider all *possible* influences. So even if the explanations we provide will tend to be in terms of what is happening nearby, we do not need the Locality Programme.

I now turn to the other purported example of action-at-a-distance, namely Bohm's version of the Einstein–Podolsky–Rosen thought experiment (Bohm, 1951).[25] We consider a system made up of two particles with spin one-half. The state of the system is such that the spins are in opposite directions. The two particles then move apart, and we measure their spins separately. In the most general version,

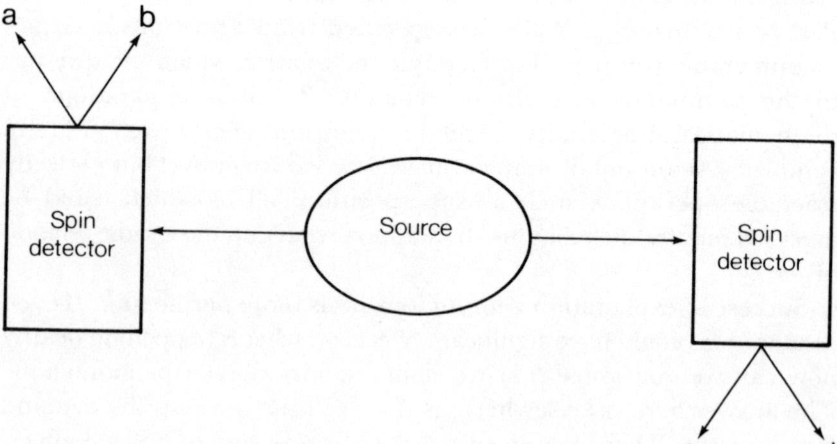

Figure 3.4 The detection of correlated spins.

we select four directions *a*, *b*, *c* and *d*, and we arrange to be able to measure the spin of the first particle either with *a* or with *b* as the straight-up direction. Likewise for the second particle, the possible directions are *c* or *d* (see figure 3.4). If we like, we may imagine levers on the spin detectors which can be put in the directions corresponding to the straight-up directions.

The initial idea behind this version of the Einstein–Podolsky–Rosen thought experiment was to provide (an idealized) method of measuring the spin-component in these directions. Suppose we observe whether the first particle is spin-up with respect to *a*, and whether the second is spin-up with respect to *c*. Say we get a Yes and a No result. Then we would be tempted to infer that, after the measurements, the first particle is spin-up with respect to both *a* and *c*. The reasoning would be that the particles have opposite spins. So a No result for the second is just like a Yes result for the first.

As for the Einstein–Podolsky–Rosen thought experiment itself, we can avoid this conclusion by appealing to something like action-at-a-distance. Current interest in it derives from the further claim that we not only *may*, but *must* accept something like action-at-a-distance. There is a statistical correlation between the results of measuring the spin of the two particles which is curiously hard to account for. This correlation is exhibited by Bell's Inequality (1964).

To state the problem which this correlation presents, we need to consider Strong Locality, which I shall now expound in an informal fashion. Later I shall present it formally. First, consider the *obvious* source of a correlation between the results of measuring the spin of two particles. It is that they start off with opposite spins. So, of course, the spins should be correlated. (To deny that would be to insist that measuring the spin not merely affects the spin, but completely randomizes it.) This suggests that we consider the state *u* of a given pair of particles, in which each particle has a determinate spin direction and they are opposite. This is a *hidden* state in the sense that it is not more directly describable using the formalism of quantum theory. So it is not a quantum state.

The spins are opposite and specified by *u*. We would expect the probability of a Yes result for a spin YNO to depend only on *u*, and not on either the choice of or the result of a spin YNO for the other. If this holds, then we say the state *u* satisfies Strong Locality. If not, it *looks* as if *either* the choice of *or* the result of the YNO for the second particle is acting directly at a distance on the first particle.

If Strong Locality failed for this state u, then we might speculate that the particles had some further structure or some further properties that were affecting the spin results. For example, a given pair of particles might be characterized not merely by opposite spins, but by having, say, the same value of another quantity, hitherto unknown to physicists. And the value of this might affect the spin measurements, preventing Strong Locality holding for u. But then we would expect Strong Locality to hold for an even more hidden state v, in which these hitherto unknown quantities are specified. What we might expect to be the case is that when we had given the *total state* of the pair of particles, that is put in all its hidden properties, then we would obtain a state s for which Strong Locality holds.[26] Then any given quantum state would be a suitable mixture of such hidden states.

Strong Locality can be stated informally by considering any two YNOs A and B about two particles distant from each other. Then if Strong Locality holds for a state s (quantum or hidden), the proportion of Yes results for A in a large collection of systems in the state s should be (approximately) the same among those repetitions for which we have Yes results for B as for those for which we have No results, and the same as for those in which B is not even observed.

More formally, a state s for a system of two separated particles satisfies *Strong Locality* if for any YNOs A_1 and A_2, for the particles 1 and 2 respectively, the results are probabilistically independent. Let Yes(X) be the proposition that YNO X has a Yes result. Let Es be the proposition that the two-particle system is in state s. Then Strong Locality states that:[27]

$$\Pr[\text{Yes}(A_1)\ \&\ \text{Yes}(A_2)/Es]$$
$$= \Pr[\text{Yes}(A_1)/Es] \times \Pr[\text{Yes}(A_2)/Es]$$

It might seem then that either there is action-at-a-distance or every state is a mixture of states for which Strong Locality holds. Now, Strong Locality will hold for decomposable pure states of two-particle systems. But it does not, in general, hold for the non-decomposable states (see chapter 2, section IV). This suggests that we should try treating the quantum state as itself a mixture of hidden states. And if Strong Locality could be restored in this way, then this might be taken as support for either a 'hidden state' speculation or a rival 'hidden variables' theory.

It is here that the Bell Inequality is relevant. It is widely held that if locality could be restored in this way then the Bell Inequality would be satisfied. But according to quantum theory it is not satisfied. Furthermore, the experimental results, with one or two exceptions, agree here with quantum theory (Redhead, 1987, pp. 107–13).

The Bell Inequality is important. But precisely what does it show? One possible approach is clearly to distinguish a number of different conjectures concerning locality, and to show that some but not others are incompatible with the formalism of quantum theory plus various 'interpretations'. For this sort of approach I refer the reader to Redhead's work (Redhead, 1983, 1987), and also to the paper by Jarrett (1984).

My approach will be somewhat different. It will, for a start, be more metaphysical and less mathematical. What I shall do is rapidly go through the derivation of the Bell Inequality, relying on Eberhard's improvement (1977), and van Fraassen's exposition (1982). This will provide an argument to show that the appeal to hidden variables will not restore locality. If this argument is endorsed then *one* reason for positing hidden variables has been found wanting. This would not, however, be a positive argument against hidden variables.

After that I discuss one of the assumptions made in the argument and show how it could be queried. However, I do eventually endorse the argument.

Finally I consider what I take to be the most significant consequence of the violation of the Bell Inequality. This is that there are non-accidental correlations at a distance (between simultaneous events) which cannot be traced back to an earlier correlation between events not at a distance. My primary interest in the Bell Inequality is precisely this puzzling feature.

First, then, let us derive the Bell Inequality. We are to consider a two-particle system in an eigenstate u for the YNO of whether the particles have opposite spins. Let $A_{i,a}$ be the YNO of whether particle number i has spin-up with respect to direction a. Then since the spins are opposite for any direction a:

(i) $\Pr[\mathrm{Yes}(A_{1,a}) \,\&\, \mathrm{Yes}(A_{2,a})/Eu] = 0.$

Perhaps u, although a precise *quantum* state, is itself a mixture of hidden states. For simplicity, suppose it is a mixture of finitely many hidden states, $s_1, \ldots, s_k$. We assume that the mixture is in positive proportions, $\lambda_1, \ldots, \lambda_k$. Then:

(ii) $\Pr[\text{Yes}(A_{1,a}) \,\&\, \text{Yes}(A_{2,a})/Eu] =$
$\lambda_1 \Pr[\text{Yes}(A_{1,a}) \,\&\, \text{Yes}(A_{2,a})/Es_1] + \ldots +$
$\lambda_k \Pr[\text{Yes}(A_{1,a}) \,\&\, \text{Yes}(A_{2,a})/Es_k]$.

From (i) and (ii), for all $j = 1 \ldots k$, we have:

(iii) $\Pr[\text{Yes}(A_{1,a}) \,\&\, \text{Yes}(A_{2,a})/Es_j] = 0$.

Let us now concentrate on just one of the s_j, and call it s. So from (iii), by Strong Locality we have:

(iv) $\Pr[\text{Yes}(A_{1,a})/Es] \times \Pr[\text{Yes}(A_{2,a})/Es] = 0$.

So either $\Pr[\text{Yes}(A_{1,a})/Es] = 0$ or $\Pr[\text{Yes}(A_{2,a})/Es] = 0$. Let b be the direction opposite a. Then $\Pr[\text{Yes}(A_{2,a})/Es] = \Pr[\text{Yes}(A_{1,b})/Es] = 1 - \Pr[\text{Yes}(A_{1,a})/Es]$.

So either $\Pr[\text{Yes}(A_{1,a})/Es] = 1$ or $\Pr[\text{Yes}(A_{1,b})/Es] = 1$.

Even if we do not assume Weak Veridicality, we could consider a characteristic of the hidden state s, that of ensuring a spin-up result with respect to direction a with probability one. Let us call a state with that characteristic an a-up state. If b is opposite to a, call a b-up state an a-down state. Then every hidden state in the mixture is either a-up or a-down.

Now suppose we have a large number of systems, numbered $1,2, \ldots N$, all in the same *quantum* state u, but not necessarily all in the same *hidden* state. We select four directions a, b, c and d. We decide to measure one of the following YNOs for each system:

$A_{1,a} \,\&\, A_{2,c}$; $A_{1,a} \,\&\, A_{2,d}$; $A_{1,b} \,\&\, A_{2,c}$ and $A_{1,b} \,\&\, A_{2,d}$.

And we select which to measure using a procedure which we take to be independent of the hidden states. That such a procedure is possible is an extra assumption. But I believe it to be quite innocuous. Denying it involves appeal to a cosmic conspiracy of such proportions that we might just as easily say that all the evidence for quantum theory is merely the (incredibly unlikely) result of chance fluctuations of the atoms in the void. I now follow the Eberhard procedure. Let $\lambda_n(x)$

take the value $+1$ if the hidden state of the nth system is x-up. Let it take the value -1 if the hidden state is x-down. Then the correlation between being x-up and being y-up is the average of the products of $\lambda_n(x)$ and $\lambda_n(y)$, that is $1/N(\lambda_1(x) \times \lambda_1(y) + \ldots + \lambda_N(x) \times \lambda_N(y)$. Call this $R(x,y)$. Eberhard considers a quantity which I call E:

$$E = R(a,c) + R(a,d) + R(b,c) - R(b,d).$$

It is easy to calculate, without appeal to any physical theory, that $E \leq 2$.[28] This is Eberhard's version of Bell's Inequality.

But we may also take a large sample of the systems, and calculate a correlation coefficient $r(x,y)$ for the Yes results for observing $A_{1,x}$ and $A_{2,y}$. Since s is either x-up or x-down, $r(x,y)$ is just the average, for the samples, of $\lambda_n(x) \times \lambda_n(y)$. Since the choice of what to measure is independent of the hidden state, we shall, except with extreme bad luck, find that the average of $\lambda_n(x) \times \lambda_n(y)$ for the sample approximates that for the whole collection of systems. Hence $r(x,y)$ will approximate $R(x,y)$. Hence E should approximate e, where $e = r(a,c) + r(c,d) + r(b,c) - r(a,d)$. But $E \leq 2$. So, except by extreme bad luck, e should be not much greater than 2. Now e is defined in terms of the correlation of the results of observations, and so is predicted by the formalism of quantum theory. It so happens that for a suitable choice of a, b, c and d, the quantity e takes values as large as 2½ (Bell, 1964). So if hidden variables in fact restore Strong Locality, the correctness of quantum theoretic predictions, even the approximate correctness, would be due to extreme bad luck. I exclude such appeals to bad luck. So I conclude that appeal to hidden variables cannot restore Strong Locality, unless some assumption we have made is rejected.

In the above we *have* made an assumption that should be scrutinized. In order to show that

$$Pr[Yes(A_{1,a}) \,\&\, Yes(A_{2,a})/Es] = 0$$

we assumed that u is the result of mixing the hidden states in *positive* proportions, and that the probabilities $Pr[Yes(A_{1,a}) \,\&\, Yes(A_{2,a})/Es]$ are *non-negative*. The latter assumption is secure enough, for these are probabilities of results of observations and so must approximate relative frequencies, which will be positive. But we should investigate the possibility that a quantum state is obtained by mixing hidden states in both positive and negative proportions. For then we should mix the probabilities in the those proportions, and could easily achieve a zero result for Pr[Yes $(A_{1,a})$ & Yes $(A_{2,a})/Eu$] without a zero result for Pr[Yes $(A_{1,a})$ & Yes $(A_{2,a})/Es$] for the hidden state *s*.

I shall not assume that such negative proportions count as negative probabilities, but that is largely a matter of stipulation. And I shall not discuss negative probability until chapter 4. But I shall consider negative proportions here, in order to complete my discussion of action-at-a-distance.

First consider the mixing of *colours*.[29] We can mix coloured light in various proportions. And since there are only two dimensions to the array of colours (ignoring the dimension of brightness) it should be possible to obtain any colour light as a mixture of three primary colours, which we may take to be pure blue, B, pure red, R, and pure green, G. But some colours turn out to have negative proportions of one of the primary colours.[30] What does this mean? Suppose the colour is $\alpha B + \beta G + \gamma R$, where α, β and γ are proportions whose sum is one. Then if, say, α is negative, we may mix $(\alpha B + \beta G + \gamma R)$ back with B in the proportions $1/(1 - \alpha)$ to $-\alpha/(1 - \alpha)$, obtaining a positive mixture of G and R.

Likewise we can try to extend the concept of mixture in quantum theory to negative proportions. For simplicity consider just two states. Our problem is that of interpreting the mixture $(1 + \lambda)u - \lambda v$, where λ is positive. This is precisely the state which when mixed with v in the positive proportions $1/(1 + \lambda)$ to $\lambda/(1 + \lambda)$ gives us u again.

We *can* mix colours in positive and negative proportions. But *can* quantum states be mixed in that way? Not if the mixture is interpreted either literally or in terms of ignorance. However, if mixture is taken to be primitive and unanalysable, then there is no reason whatever for refusing to extend mixture to negative proportions. We have, therefore, a possible way out. We could explain those curious statistical correlations (the violations of Strong Locality) by saying that they follow from the way quantum states are mixtures in positive and negative proportions of hidden states for which locality would hold. Notice, however, that since this is not a literal mixture those hidden states would be merely possible ones. They would not actually occur.

If this were the only reason for taking mixture to be unanalysable, then it would not be worth it. The mystery of unanalysability equals or even exceeds that of an unexplained correlation at a distance.

It would be a sensible explanation only if we were otherwise convinced that mixture is unanalysable.

In my discussion of mixture I argued that the statistical interpretation was the appropriate one if we hold a Determinate Particle Speculation. And it is only on *that* speculation that the failure of the Strong Locality requirement implies a correlation at a distance. (This will be discussed further in chapters 6 and 7. But roughly if the 'particles' are all over the place, the correlation is *not* at a distance.) The upshot is that I have no use for the proposal that we preserve Strong Locality by treating mixture as primitive and unanalysable. So I reach the standard conclusion: On the Determinate Particle Speculation there are unexplained correlations at a distance.

Has Bell's Inequality shown us anything we could not have concluded from the twin-slit thought experiment? It has, because of the symmetry of the set-up for Bell's Inequality. In the twin-slit experiment it is natural to think of the arrangement of slits as both preceding and causing the behaviour of the particles. So we might conclude that there is action-at-a-distance, but not simultaneous action-at-a-distance. By contrast, in the set-up for Bell's Inequality we may assume that the measurements are made simultaneously. (Or in a relativistic context we may assume that neither is earlier than the other.) In this case then, we have a correlation at a distance between simultaneous events. *That* is the new feature of the situation.

It is easy to jump to the conclusion that we could send a message on the 'Bell' telephone, or in some other way *use* the choice of YNO for one system to affect the result of measurement of the other. This, combined with Special Relativity, would lead to paradoxes.[31] Thus we would have a reductio ad absurdum of the conjunction of Quantum Theory with Special Relativity.

Fortunately, nothing so drastic follows. I know of no argument to show that the choice of which YNO to measure for one particle *affects* the other particle.[32] Hence the 'Bell' telephone is like that other excellent invention, the perpetual motion machine.

What are we to say, then? Is it that the *result* of one YNO, rather than the choice of which to make, affects the other particle? The symmetry of the situation makes this implausible. Why should making a YNO on one particle be picked out as cause rather than effect?

What we should say is that the results of the YNOs are correlated in a non-accidental way *without any causal explanation* for this correlation. This is, in turn, taken by van Fraassen as a refutation of what he calls *epistemic realism*, the position that 'Reasonable expectation of future events is possible only on the basis of some understanding of (or, reasonable certainty about) causal mechanisms that produce these events' (van Fraassen, 1982, p. 26).

Since these non-accidental statistical correlations do enable us to have some reasonable expectations about the future, namely future correlations, I agree with van Fraassen. I am a scientific realist, but not an epistemic realist in van Fraassen's sense.

We have argued that there are non-accidental correlations between causally unrelated events. Here is the genuinely puzzling novelty which *this* purported case of action-at-a-distance involved. Now I think our current concept of a law of nature is a careless secularization of the concept of a decree by God – one suited neither for religious nor secular purposes (Forrest, 1985b). But these worries are not here relevant. So I shall call these non-accidental correlations *laws*, for short.

The lesson of Bell's Inequality is, then, that there can be law-governed statistical correlations without causation. This puzzles us. But why are we puzzled? I know of no plausible theory of laws which excludes them (see, for example, Armstrong, 1983, pp. 135–6). I trace the puzzlement to some pervasive *models* (in the sense of *exemplars*) of explanation. The most directly relevant is the model in which we explain the occurrence of an event by showing it to be caused by some other event. Quite compatible with this, but without the emphasis on causation, is the historical model of explanation where we seek to understand something by tracing back its history. Influenced by these models we tend to assume that the *very best* explanations of events are ones that show that an event is inevitable given earlier events. And the influence of the models suggests that the laws should at least be causal ones, even when we no longer have determinism. That is, they should show that later events are the effects – albeit probabilistic ones – of earlier events.

The law-governed correlations without causation puzzle us because they do not conform to causal and historical models of explanation. What we need is another model of explanation. I call it the Limited Possibility Model. Here is an example. Suppose we are wondering why various crystals exhibit the sorts of symmetry they do. One sort of explanation would be to investigate the details of the chemical binding and show that a given substance must crystallize in a certain way. And no doubt that is the best kind of explanation. But surely some partial understanding of the situation is provided by an exercise in pure mathematics showing that the possible ways in which crystals could exhibit symmetry are fewer than we might expect. This would be an example of the Limited Possibility Model of Explanation.

Let us be guided by the Limited Possibility Model. What sort of laws would we look for? Surely we would look for various *constraints*. Examples of such constraints are conservation laws. They do not tell us what causes what but are merely constraints on the system. Non-causal statistical correlations are, I submit, *constraints* which fit in with the Limited Possibility Model of Explanation. We explain in part by showing that fewer things were possible than we initially thought.

I shall call the laws that merely provide constraints, *negative* laws. Those that tell us what will (or is likely to) cause what, I call *positive* laws. The Limited Possibility Model of Explanation should make the occurrence of fundamental negative laws less puzzling.

To sum up: purported cases of action-at-a-distance do not threaten particle speculations. First, action-at-a-distance *as such* is nothing to worry about. *Simultaneous* action-at-a-distance is more problematic – but there is no evidence that it occurs. However, we have found some interesting results. The Dyadic Programme has to be abandoned, although Methodological Atomism need not be. Further, we have to allow that there might be non-causal law-governed simultaneous correlations between distant events.

Neither action-at-a-distance nor law-governed simultaneous correlations at a distance will be a deciding factor when comparing speculations. Because of their interest, however, I shall mention them in my comparisons.

NOTES

1 Some problems, for example those raised by the No Hidden Variables Theorems, are of significance primarily for just one 'interpretation'. I delay discussing such problems until the appropriate 'interpretation' is discussed.

2 Aesthetic Resistance Movement.

3 That is, cases in which effects are earlier than their causes.

4 Von Neumann also considers the possibility that such pure state collections are impossible to prepare and so we should consider mixed states. He argues (1955, pp. 437–9) that this will not affect the subsequent discussion.

5 For simplicity consider measurements of spin along three perpendicular axes *Ox*, *Oy* and *Oz*. Again, for simplicity, consider the idealized case

where the particle has spin but no location. Let u^+_x and u^-_x denote the eigenstates for the YNOs of whether the spin is up and down, respectively, with respect to the *Ox* axis. Now consider the mixture $1/2u^+_x + 1/2u^-_x$. The probabilities of a Yes result for the YNO of whether the spin is up with respect to the directions *Oy* and *Oz* are both near 1/2. The only pure states which even approximate these probabilities are ones which are near (i.e. make a small 'angle' with) either u^+_x or u^-_x. But for these the probability of a Yes result for the YNO of whether the spin is up with respect to the x direction is near 1 or 0 respectively. Yet for the mixture, the value is 1/2. So in this case, no pure state is even approximately indistinguishable from the mixture.

6 Interaction with some other system can result in a transition from a pure state to a mixture even if, initially, the other system is itself in a pure state. For in that case the composite system is initially in a decomposable pure state. But this may evolve into a non-decomposable pure state of the composite system (see chapter 2, section IV). Unless we deny the Reducibility of Composite Systems (see section VI), in such a non-decomposable pure state the component systems must both be in mixed states.

7 Cartwright appeals to a formulation of statistical quantum theory (Davies, 1976) in which the group of unitary operators expressing time evolution in the standard Hilbert space formalism is replaced by a semigroup of operators which need not be unitary. However she does not indicate what procedure would govern the construction of that semigroup. It is easy enough to derive such a semigroup by combining the Schroedinger and von Neumann dynamics. But that is not what is required if these are to be seen as special cases. Cartwright's proposal remains an interesting *suggestion* at this stage.

8 It is common practice to use the names of great philosophers as labels for positions based on, or reminiscent of, ones they have held, without regard for scholarship. (For example Platonism often means belief in uninstantiated universals, but Plato held that the Forms were self-instantiated.) In the same spirit I use the phrase 'Kantian speculation' for something less subtle than Kant's critical philosophy.

9 The predicates 'grue' and 'bleen' were introduced into philosophical discourse by Goodman (1954). I take 'grue' simply to mean 'green before 2000 AD, blue after 2000 AD', and vice versa for 'bleen'.

10 Lewis (1983, p. 347) says that an *adequate* theory of properties is 'one that recognises an objective difference between natural and unnatural properties; preferably, a difference that admits of degrees'. I here assume a theory of properties which is adequate in Lewis's sense. And I require that a scientific hypothesis be based on a classification of data into types in such a way that the types are picked out by properties of a high degree of naturalness.

11 *Suitable* YNOs include position and momentum YNOs for *convex* regions U of space, or *convex* sets of momenta. The restriction to convex sets is required if we are to take Wigner's formula for the joint distribution of position and momentum seriously. For this formula is:

$$\Pr(q,p) = K\int \overline{\Psi}(q+x)\Psi(q-x)e^{2ipx}\,dx.$$

An eigenstate for a region U is represented by function Ψ zero outside U. So $\Pr(q,p)$ will be zero unless for some x both $q+x$ and $q-x$ are in U. If U is convex this ensures that q is in U. So the eigenstates are given by distributions which are zero outside U. And similarly for momentum. This would not hold if the regions were not convex.

12 This is called the *eigenvalue–eigenstate* link by Fine (1973a), who rejects it.

13 Based on the diffraction of β rays.

14 This is an instance of the Radical Privilege (see chapter 4, section II).

15 See note 33 of chapter 2.

16 Let the initial mid-point of the two particles be O. Select some small positive real number δ. Then we may divide space up into countably many disjoint cubes C_1, C_2, etc. of volume δ. Let the cube D_i be diametrically opposite C_i. (So O is midway between C_i and D_i.) Then any eigenstate for the YNO of whether the mid-point is O will be represented by a vector $\Sigma\alpha_i x_i \otimes y_i$. Here x_i represents an eigenstate for the YNO of whether the particle is in C_i, and y_i represents an eigenstate for the YNO of whether the particle is in D_i. The vectors $x_i \otimes y_i$ are orthogonal. So if we now discover which of the cubes C_i the first particle is in, the von Neumann dynamics ensures that the resulting state is represented by $x_i \otimes y_i$ for some i. This is still an eigenstate for the YNO of whether the mid-point is O.

17 The hydrogen atom is in a superposition of decomposable states. If the electron is in pure state x_o, then each of the decomposable states is $x_o \times y_i$ for various y_i. But then the hydrogen atom is itself in a state $x_o \times y$ where y is a suitable superposition of the y_i.

18 I do not here assume any particular formula for obtaining the state of a subsystem if we know the state of the system as a whole. The standard formula (known as the Reduction Formula) is not required for this discussion. For a statement of it see van Fraassen (1973, p. 295).

19 If u and v are represented by the one-dimensional subspaces spanned by unit vectors x and y respectively, then we may take w and z to be represented by the one-dimensional subspaces spanned by $(1/\sqrt{2})(x \pm y)$. The mixed states $1/2\,u + 1/2\,v$ and $1/2\,w + 1/2\,z$ are then represented by the operators T_1 and T_2 where, for any vector s in the Hilbert space:

$$T_1(s) = 1/2\langle s|x\rangle x + 1/2\langle s|y\rangle y$$

$$T_2(s) = 1/2\langle s|(1/\sqrt{2})(x+y)\rangle(1/\sqrt{2})(x+y) + 1/2\langle s|(1/\sqrt{2})(x-y)\rangle(1/\sqrt{2})(x-y)$$

By some elementary algebra $T_1(s) = T_2(s)$. But s is any vector in the Hilbert space. So $T_1 = T_2$.

20 The appropriate Hilbert space is two-dimensional. Hence the mixture of any pair of orthogonal states in equal proportions turns out to be ½ I, where I is the identity operator. Here I assume that we have no spin/location or spin/momentum correlations. These would complicate matters.

21 By *directly* I mean without causal intermediary, rather than without lapse of time.

22 By Cliff Hooker (private communication).

23 On the Wave and on the Indeterminate Particle Speculations, there are total and associated energies, respectively, which are equal to the mean values of the energy specified by the probabilities. These too are conserved. If we give a statistical account of the probabilities in quantum theory as frequencies, then we might conjecture that the energy is conserved for individual systems in the collection. Although this conjecture is not forced upon us by the conservation of probability-distributions for energy, it is suggested by an explanation of the scattering of light by electrons (the Compton effect). For the scattering can be explained by energy and momentum conservation (Das and Melissinos, 1986, pp. 9–10). However, the tunnel effect, to be discussed in chapter 5, section 9 strongly suggests that energy can be 'borrowed' for a short enough time. In that case the conjecture needs qualification.

24 Ignoring spin, the energy *operator* is the same (Hamiltonian) function of the position and momentum *operators* as the energy in classical mechanics is a function of position and momentum. Even allowing for spin we obtain a classical-type formula. And the position, momentum and energy operators are themselves suitable integrals of the projection operators representing position, momentum and energy YNOs. For example the position operator Q for a single particle is $\int q\,\mathrm{dM}$, where the 'measure' M assigns to the region V the operator representing the YNO of whether the particle is in region V.

25 For a detailed account of the Einstein–Podolsky–Rosen experiment and Bell's Inequality see Redhead (1987, chs 3, 4).

26 This would be expected even if there is an infinite regress of structure in which particles are made of particles, which are made of particles, without end. In that case, the hidden state would not be describable, being infinitely complex. But this does not affect the discussion.

27 Compare Jarrett (1984, p. 581). I call this condition *Strong Locality*, because, apart from the change of notation, and apart from any underlying differences expressed by the choice of notation, this is Jarrett's Strong Locality.

28 $E = 1/N(E_1 + \ldots + E_N)$ where $E_n = \lambda_n(a) \times \lambda_n(c) + \lambda_n(a) \times \lambda_n(d) + \lambda_n(b) \times \lambda_n(c) - \lambda_n(b) \times \lambda_n(d)$.

So $E_n = \lambda_n(a) \times [\lambda_n(c) + \lambda_n(d)] + \lambda_n(b) [\lambda_n(c) - \lambda_n(a)]$.

Therefore $|E_n| \leq |\lambda_n(a)| \times |(\lambda_n(c) + \lambda_n(a)| + |\lambda_n(b)| \times |\lambda_n(c) - \lambda_n(d)|$

Each $\lambda_n(x)$ is either $+1$ or -1 so either $|\lambda_n(c) + \lambda_n(d)| = 2$ and $|\lambda_n(c) - \lambda_n(d)| = 0$ or $|\lambda_n(c) - \lambda_n(d)| = 2$ and $|\lambda_n(c) + \lambda_n(d)| = 0$.

In the first case $|E_n| \leq 2|\lambda_n(a)| = 2$.

Likewise in the second $|E_n| \leq 2|\lambda_n(b)| = 2$.

Hence $|E| \leq 2$.

29 A similar point could be made by considering polarized light.

30 Given one choice of standard pure blue, pure green and pure red, wavelengths from 380 to 545 have negative red and those from 530 to 670 negative blue (Wyszecki and Stiles, 1967, table 3.5).

31 The basic idea is that it would then be possible for person *A* to send a message to *B* which arrived earlier than it was sent (with respect to a suitable frame of reference). Then *B* could send a message back to *A* which likewise arrived earlier than it was sent. Combining this with common-sense assumptions about what messages people are free to send, we could get paradoxes in which *A*'s message depends on *B*'s and vice versa. For a detailed discussion see Redhead (1983, pp. 166–70).

32 That would require violation of a weaker condition, namely *Locality* in Jarrett's sense. Jarrett shows that Strong Locality is equivalent to the conjunction of Locality and what he calls *Completeness* (Jarrett, 1984). His conclusion is that Strong Locality is violated, but that Locality is not. For another analysis of the situation, which also supports the conclusion that the 'Bell' telephone cannot work, see Ghirardi et al. (1980).

4

The Copenhagen School of Metaphysics

Thinkers such as Bohr, Heisenberg and Born[1] who made up the Copenhagen School provided us both with a formulation of quantum theory and with various pieces of 'interpretation', that is metaphysical speculation. Talk of a single Copenhagen Interpretation is misleading, unless by that you mean simply the formulation of quantum theory which became the orthodox one. In this chapter I am concerned not with the formulation but with the metaphysical ideas that these thinkers, especially Bohr and Heisenberg, produced. I suspect that there was considerable disagreement among them on matters of metaphysics and that their views changed with time. But my purpose is not exegetical. It is to discuss some important metaphysical ideas which arose from the Copenhagen School, criticizing some and endorsing others.

I INDETERMINACY AND THE RIGHT TO CONCEPTUAL INNOVATION

In classical mechanics a particle is considered to have a determinate location at all times. Even if it is not a point particle, by taking the location to be the centre of mass, we may take the location to be a single point. Because a classical particle has a determinate location at all times (and because it moves smoothly) we can *define* its velocity as the rate of change of location. So, assuming the particle has a precise mass, we could define its occurrent momentum as the mass times the velocity. Alternatively we could posit a dispositional momentum, namely a disposition to change location. In that case momentum would be *manifested* by a velocity equal to the numerical value assigned to the momentum divided by the mass. In either case a classical particle has a determinate momentum.

One of the ideas that comes out of the Copenhagen School is that particles might have indeterminate location (and indeterminate momentum, but I shall concentrate on location). The particles concerned are assumed to have a radius no greater than some specified value, and lack of determinateness in the position far exceeds this. So it is not simply that particles are extended. The indeterminacy of location results in a failure of bivalence. For a particle whose location is in region *V* but which has no more determinate location, the statement that it is in some region *U* contained in *V* is neither true nor false. Such a failure of bivalence is, as I have explained in chapter 1, quite compatible with realism about particles in my sense.

Now I shall argue in chapter 7 that one tenable speculation involves particles of indeterminate location. But I believe there is also a tenable speculation in which particles always have determinate locations. What I take from the Copenhagen School, and what I shall now defend, is not, therefore, a *duty*, but a *right* to conceptual innovation and, in particular, the right to posit particles of indeterminate location.

Audi (1973, pp. 31–2) has provided a representative argument against such conceptual innovation. He quotes from a standard text that quantum mechanics 'contains classical mechanics as a limiting case, yet at the same time it requires this limiting case for its own formulation' (Landau and Lifschitz, 1958, pp. 2–3). The objection is that when describing the measuring apparatus we assume there are particles with determinate position, yet when thinking of the system itself Bohr and Heisenberg deny there are such particles. The reply to that objection is fairly obvious. All we require is that the measuring apparatus is classical to a good approximation. The lack of determinateness may be posited to be negligible in such cases. But this leads to a further objection which will apply to any conceptual innovation in quantum theory. It is the threat of incommensurability between it and classical mechanics. The threat is that if I call an electron in classical mechanics a *particle* and call an electron in quantum theory a *particle* then I am equivocating on the word 'particle', and hence, arguably, on the word 'electron'. But in that case quantum theory could never be approximated by classical mechanics in the fashion I have assumed. This criticism is half of a familiar thesis of Kuhn's (1970), that wherever there has been a change of scientific paradigms we have conceptual innovation, and hence such an equivocation. However, Audi does not have to endorse all of Kuhn's thesis. For in this case the conceptual innovation is admitted, even gloried in, by members of the Copenhagen School, while in general it could be denied that changes in paradigms result in a comparable change of concepts.

My reply to this objection is a crude one. But why be subtle if crudity suffices? The objection is due, I say, to equivocation about equivocity. Suppose you and I use the word 'father' somewhat differently. I assume that a father must be male. You claim that anyone who has fathered a child is still a father even though that parent's sex subsequently changed. We have different concepts of fatherhood. Does this mean that we inevitably talk at cross purposes? By no means. For in almost all cases if 'X is the father of Y' is true, as I understand it, it will be true as you understand it. So you can use what I say as a source of information in spite of our conceptual differences. To be sure there is *potential* for equivocation about the word 'father', but this potential need not be actualized. I draw the moral that conceptual innovation always introduces a potential for equivocation, but there need to be no actual failure to communicate.

In the case of determinate versus indeterminate particles actual equivocation only arises when we make remarks such as 'If it is a particle then it must have determinate location'. However, there is no talking at cross purposes if we say 'To a good approximation the particle has location q'. For that can be true of both a classical and a non-classical particle. Provided there is some truth of the form 'The particle is in region V' where V is a small region around q, then, whatever concept of a particle we have, the particle is approximately at q.

It might still be objected that although such statements about classical and non-classical particles could be guaranteed to have the same truth-values they none the less have different meanings. But recall the dialectic. The objection started off as a charge that we assume particles have determinate positions because we use classical mechanics to describe the measuring apparatus. My reply was that all we have to assume is a good approximation to classical mechanics. And I have now argued that we can have statements about non-classical particles which provide a good approximation. Further questions about meaning are irrelevant.

II THE TAINT OF POSITIVISM

By *Positivism* I mean Logical Positivism, namely the thesis that all meaningful statements are either logical (or analytic) truths, on the one hand, or can be verified directly by observation, on the other. The New Metaphysics, engaged as it is in metaphysical speculation, presupposes the rejection of Positivism. So it might seem that I should reject the Copenhagen School and all its works, for it is widely thought to be influenced by Positivism. Now I concede that Bohr and Heisenberg said some positivist-sounding things, and presumably meant them. Thus Heisenberg says: 'Any use of the words "position" and "velocity" with an accuracy exceeding that given by . . . [the uncertainty relation] is just as meaningless as the use of words whose sense is not defined' (1930, p. 15). This seems to be a positivist inference *from* the impossibility of discovering the position and the velocity with greater accuracy *to* the meaninglessness of assertions that they have such a position and velocity.

However, I think there is much good in the Copenhagen School. Rather than totally rejecting it, I reject the Positivism in it. It is notable, in this connection, that members of the School seemed to have been realists about many unobservables, hence they were never instrumentalists, as stereotype positivists would have been.[2]

Although I reject Positivism I admit that it performed a useful function in the Copenhagen School of Metaphysics. I shall now identify this function and show that we do not *need* Positivism to perform it. The radical asks: Why should the old ways of thinking be accepted because they are well entrenched? The conservative answer is that we are heirs to a tradition which we cannot fully evaluate or criticize, for it is too rich and varied. There is a presumption that there are good reasons for the tradition even if we cannot comprehend them. This conservative appeal to intellectual humility is, I think, basically sound. But it establishes no more than a presumption which can be overcome. And one way in which it can be overcome is by a version of Ockham's Razor. If we have reason to believe that there is no observable difference between *X*s and *Y*s, or that *Z*s are unobservable, then we have the privilege of repudiating a tradition which tells us *X*s are not *Y*s, or that *Z*s exist. We may exercise this privilege if there are good theoretical grounds for so doing. An example is that of simultaneity. Given Special Relativity there is no way of observing the exact simultaneity of two events. So we have the privilege of denying that there is a relation of simultaneity. Likewise with the conceptual innovations of quantum theory. The Conservative Presumption that particles always have determinate location even if the state is one of a fairly precise momentum may be overcome because, by the Uncertainty Principle, it is impossible to measure that location with any accuracy.

I shall call the privilege of overcoming the Conservative Presumption (in cases where what tradition supports can be argued to be unobservable) the Radical Privilege. Now the cases where Heisenberg and others talk positivistically are cases where all they need is to invoke the Radical Privilege. Positivism implies the Radical Privilege. Indeed its defect is that it turns an intellectual right into an intellectual duty. But we do not need Positivism in order to retain the Radical Privilege. For it can be supported in a much less controversial way. All that is required is that we be *empirical*.[3] To be empirical is in part to be suspicious of claims to immediate knowledge of momentous truths. To be empirical is, for instance, to think that any knowledge of God's existence would have to be the result of argument or experience, rather than direct intellectual insight. Being empirical goes naturally with the non-dogmatic metaphysics I am propounding. And I shall assume the reader is empirical in this sense. It follows that we need not be too impressed by claims to know something which the tradition asserts but which is unobservable. And that is all the justification which the Radical Privilege requires. The strength of tradition lies, I submit, more in the wealth of collective experience than in any accumulation of intellectual insight denied to most individuals. So in the area of the unobservable we are free to reject tradition.

What physicists need, then, is the Radical Privilege. But they do not have to endorse Positivism to obtain this privilege. However, in the case of indeterminacy it is especially tempting to do so. For what is being claimed is that various statements are neither true nor false. Now, I believe that this claim is a speculation, in which case the statements in question are not nonsense. But that is a somewhat *recherché* way of being neither true nor false. The obvious way is by being meaningless. We can see therefore why Heisenberg and others appealed to Positivism. However, this appeal is unnecessary.

III COMPLEMENTARITY

Complementarity is Bohr's baby, and other members of the Copenhagen School treat it as such. Bohr suggests (1949) that we need to consider pairs of 'pictures', neither of which is completely adequate.

Thus there is a particle 'picture' and a wave 'picture', a position 'picture' and a momentum 'picture'.

I accept something like complementarity. For I believe there are various different metaphysical accounts – I have three – which are all equally tenable. Indeed more than one may be true. In that case the thesis that each account is in some way a partial 'picture', though a vague one, seems reasonable enough. But that is not what Bohr had in mind. He seems to have been considering a single metaphysical speculation based on a supposed need to switch back and forth between rival 'pictures'. I submit that complementarity in that sense was a relic of an earlier stage in quantum theory. Before the mathematical theory was fully developed the only way to handle specific problems was to use sometimes one sometimes another 'picture'. But as Heisenberg points out (1930, pp. 10–11), once we have an adequate mathematical scheme we no longer need these 'pictures'.

Complementarity is redundant as a piece of formalism. But it could still be proposed as a metaphysical *speculation*. First consider wave/particle complementarity. If we were considering rival metaphysical speculations, then, indeed, a wave, a particle of indeterminate location and a particle with determinate location initially seem quite different sorts of speculation. So, initially at least, we have three rival theories or pictures. Do we need a mixed theory as well? The answer is 'No'. For waves can do whatever particles can do. More precisely, a wave can mimic a particle. For a wave is just any disturbance to a field, it need not be wavy (i.e. sinusoidal). So we can have waves which have minimal disturbance except in a small region (see figure 4.1). Such a wave will be very like a particle. Consequently if it were the case that we *sometimes* needed a wave 'picture' then we could *always* rely on a wave 'picture' and never have

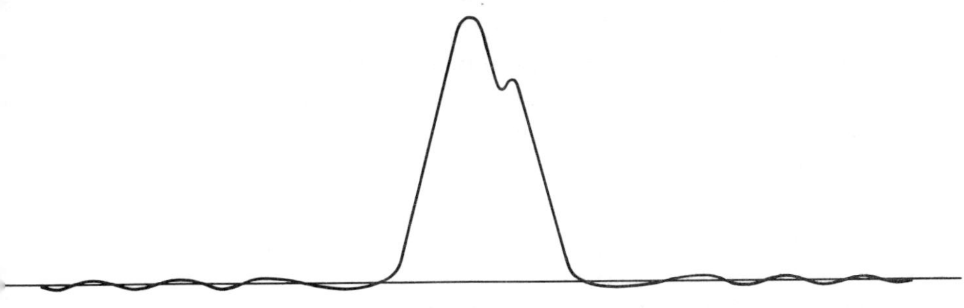

Figure 4.1 An approximately localized wave.

to resort to a particle 'picture'. Admittedly we would then be faced with the problem of deciding when a wave can spread and when it cannot. But that is a minor difficulty.

Quite generally, it only seems necessary to resort to wave/particle complementarity if there is neither a satisfactory wave speculation nor a satisfactory particle speculation. This raises the question of just *how* unsatisfactory such speculations would have to be before we should prefer switching from one to the other using wave/particle complementarity. I leave this to the reader to decide. But, in any case, Bohr's position would seem one to resort to only if all else fails.

Again, position/momentum complementarity was proposed. In this case the two 'pictures' were a position 'picture', in which a particle has determinate position but no determinate momentum, and a momentum 'picture', in which a particle has determinate momentum but no determinate position. This is not tenable. For the position 'picture' was applied to those cases where the variance for observation of position is small. Likewise, the momentum 'picture' was applied to those cases where the variance for observation of momentum is small. But there is a continuum of intermediate cases.[4] If we say that neither 'picture' applies in such cases, then the theory of position/momentum complementarity is inadequate. If we say both apply in such cases, then we should say that both apply in all cases.

In reply to this, it could be pointed out that on Bohr's account (Hooker, 1972, pp. 133–5) the 'pictures' are not used at random. We are to use whichever is appropriate for the kinds of measurement being made. In the intermediate cases we are able to measure neither position nor momentum with precision. But both are measurable *approximately*. This amounts to saying that in some situations neither a determinate position nor a determinate momentum 'picture' is appropriate, but rather a picture with indeterminate position and indeterminate momentum. My rejoinder to this reply is that we now have a single 'picture' once more, namely the Indeterminate Particle Speculation, according to which, in general, particles have indeterminate positions and indeterminate momenta, and determinacy is a very special case. I develop this in chapter 7. And I acknowledge that this is a successor theory to some of the Copenhagen speculations, including position/momentum complementarity. Likewise if we should accept the Feyerabend–Hooker argument (Hooker, 1971) that there cannot be both determinate energies and determinate time, we need not resort to a Bohrian complementarity of energy and time but rather to the successor theory – the Indeterminate Particle Speculation – according to some versions of which there are indeterminate energies.[5]

Thus far I have stressed the greater intellectual economy of single 'picture' speculations. But there is a further criticism of Bohr's position. If he *merely* said that sometimes one 'picture' sometimes another was applicable, this proposal would be ad hoc. Instead, he gave us an account of *which* 'picture' is appropriate in *which* circumstances. It is the kind of observation which specifies the 'picture'. As Hooker points out (1972, pp. 193–208), this leads to a somewhat Kantian position. By requiring the description of a system to be in terms provided by one or another 'picture', and then inisting that the 'picture' to be used depends on the way we observe the system, Bohr prohibits the description of unobserved systems. This amounts to denying that we can know things as they are in themselves, and to insisting that we only know things as they are related to us in observation. Hence Bohr's position is subject to all the criticisms of the Kantian position, including those I made in the previous chapter.

I conclude that we should reject complementarity as a speculation, but I acknowledge the influence of Bohr's position on the Indeterminate Particle Speculation which I defend in chapter 7. In addition, the Pluralist Presumption – to be discussed in chapter 8 – is somewhat similar to Complementarity. For although Complementarity is not tenable as a single speculation, we might still consider the possibility of a complementarity between speculations. Consider wave/particle complementarity. By the Pluralist Presumption we might think we could accept both a wave and a particle speculation. I shall discuss this in chapter 8.

IV PROPENSITIES

The state of a quantum system specifies various probabilities for the results of the various YNOs. For example, in the case of a single particle it specifies, among other things, a probability for a Yes result to the YNO of whether the particle is in some region V. Whatever else these probabilities may be, they are *usually* manifested by the frequencies of the occurrence of Yes results if the same YNO is asked for a large collection of systems. The idea of a propensity to ϕ is roughly the idea of a property X of a single system which is usually manifested by the frequencies with which systems with this property X actually ϕ

in a large collection of such systems. Propensities are somewhat analogous to dispositions. A disposition to ϕ is a property manifested by something ϕ-ing in the appropriate circumstances. Something can have an unmanifested disposition. (A brittle vase need never shatter if it is never dropped.) In a similar fashion a high propensity to ϕ is not *necessarily* manifested by a high frequency of ϕ-ing, for there may not in fact be a large collection, or, by chance, only a few of the systems might ϕ.[6] That is why I say that the propensities are *usually* manifested by the frequencies.

This idea of a propensity has been championed by Popper (1982, pp. 64–74), without whom it would not have been widely considered by philosophers. However it can be traced back to the Copenhagen School. Born, for instance, says 'I personally like to regard a probability wave . . . as a real thing, certainly as more than a tool for mathematical calculations' (1949, pp. 105–6). I interpret this statement as the claim that some of the probabilities mentioned in quantum theory are to be treated as propensities about which Born is a realist. Again Heisenberg (1959, p. 53) suggests that 'the "probability function" contains statements about possibilities or better, tendencies ("potentia" in Aristotelian philosophy)'.

This is an appropriate place to discuss propensities in greater detail. Propensities may seem to be rather mysterious. What I aim to do is to provide *one* non-mysterious account of propensities. At least, I aim for an account which is no more mysterious than certain laws of nature. There may well be other accounts: mine has the advantage that it permits negative propensities. If the reader has some alternative account which does not, then I claim this as an advantage for my account. Otherwise, the reader may substitute his own account for mine. In short, my purpose is not to find all satisfactory accounts of propensities, but rather to justify my own subsequent reliance on propensities by exhibiting one such account.

Propensities, like dispositions, are often contrasted with *occurrent* properties. Prima facie the distinction here is between a property like brittleness which gets manifested in some way (such as by the vase breaking) and a property such as being of mass one kilogram which the object just *has*. But I prefer an alternative. I shall not distinguish two sorts of properties, but rather distinguish between two sorts of ways of *describing* or *knowing* properties. Thus having mass one kilogram could be described non-occurrently as the property of exerting such and such a pressure on an object supporting something with that property. Conversely, we might follow Armstrong (1973, pp. 14–15) in identifying the property of brittleness with a certain occurrent property, namely the occurrent property in virtue of which the object will smash if dropped.[7] To avoid positing special kinds of properties, I shall follow Armstrong and treat propensities as properties described in a special way.[8]

One suggestion is that a propensity of degree α to ϕ is just some property F-ness such that in a large collection of objects with F-ness, a proportion near to α will ϕ. A minor difficulty with this is the vagueness of the 'near to' qualification. Much more serious is the small, but non-zero, probability that even in a large collection a proportion not near to α will ϕ. Even if α is near one, by chance, nothing with the propensity of degree α to ϕ might ϕ. For this reason we must characterize propensities another way and connect them with frequencies less directly.

Following Mellor (1971) and Lewis (1979), I relate propensities to *degrees of belief*, or *subjective probabilities* as they are often called. I assume that we can *represent* various *degrees* of belief, agnosticism and disbelief, on a numerical scale from 0 to 1.[9] Subjective certainty or full belief is represented by 1. Full disbelief is represented by 0. And perfect agnosticism by 1/2. Now suppose someone had some specified background information. Suppose further that she comes to know that some item x is F, where being F is an intrinsic property of x. Finally suppose she has no further information, other than this and the specified background information. If in those circumstances the one and only warranted degree of belief that x will ϕ is of degree α, then we may say that the property of *being* F just is the propensity of degree α to ϕ. On this account propensities are relative to some background information. And, for that reason, the concept of a propensity, at least as I have analysed it, is only useful if there is some especially *salient* background information.

What *is* the background information that we need to consider in the present context? It should be such as indeed warrants the degree of belief. And the only non-mysterious information of this sort would seem to be *statistical* information about relative frequencies. To avoid the problem that by chance the relative frequencies could depart from propensities I propose the following hypothesis, clauses (iii) and (iv) of which I take to be the background information. For the sake of definiteness, I shall state this for the case where it is a quantum *system* which has the propensity. But we may substitute 'particle' for 'system' if we are considering the propensities of *particles*.

The hypothesis is that:

(i) The system has various intrinsic properties (the propensities).
(ii) For each such property we can consider large collections of systems with that property. If the corresponding frequencies approximate those specified in the description of that property as a propensity, I say the collection *conforms* to the propensities of its members.
(iii) The proportion of large collections conforming to the propensities of their members is very high.[10]

and:

(iv) That the proportion of conforming collections is very high is itself no accident but is distinguished in whatever way the so-called laws of nature are to be distinguished from merely accidental regularities.

Clause (iv) tells us that if, for instance, you believe there are no laws of nature strictly speaking (Forrest, 1985a), but that some regularities are less accidental than others, then you are to say that the statistical constraints mentioned in (iii) are some of the less accidental ones. Or if you hold that laws of nature are relations between universals, then the statistical constraints should also be laws reflecting relations between universals.[11] And similarly for other accounts of laws of nature.

Notice that this account is holistic in the sense that (iii) does not entail that a high proportion of large collections for *one* specified set of propensities conform, only that a large proportion of *all* large collections conform. None the less, because of (iv), a propensity theory supports counterfactual conditionals of the sort: If there were to be a large collection of a certain kind it would very probably conform.

Before proceeding, I shall consider three objections to my account of propensities. The first is that, so the objector says, the proportion of large collections which conform to the propensity, could, by chance, be low. While initially this looks just like the original objection to a frequency account of propensity, it is not. For we can *argue* that it is likely that somewhere, some time, frequencies in a large collection will not approximate the degree of the propensities. However, there is only *one* collection of all large collections. And it is part of my hypothesis, namely (iii), that in this collection the proportion of conforming collections is high. And, by (iv), that it is high is due to the laws of nature. The objection derives a spurious force from an illegitimate extrapolation from individual collections to the collection of all collections.

The second objection is that I am being sceptical when I insist that propensities are not special kinds of properties, but properties described in a special way. The objector asks: Why should there not be special properties which give things powers or tendencies to cause various effects? What Popper says can be read that way. He suggests that propensities be interpreted as 'tendencies to realize themselves' (1982, p. 129), and he says that 'they lead to corresponding statistical frequencies' (pp. 129–30). This certainly suggests that the statistical constraints are to be *explained* by the propensities, whereas on the account I have stated, the properties are only described as propensities because of constraints on the statistics. I say there is no need to posit such special properties. So we should not do so.

The third objection is the opposite of the second. It is that my talk of propensities is redundant. Why not short-circuit it and talk just in terms of constraints on statistics? There are three reasons. The first is that, by talking in terms of propensities, I enable the reader who disagrees with me about what propensities are to make the necessary substitutions with ease. The second is that the constraint on the statistics is of a global or holistic sort. And it is convenient to have some way of describing the probabilistic constraint this puts on the individual systems. Talk of propensities does this. Thirdly, it is convenient to have a reminder that various intrinsic properties of the systems being considered (the ones we have described as the propensities) play a special role in the statistical constraints.

In the account I have given, a propensity of degree α to ϕ is *manifested* by a large collection of items each with that propensity. And in the vast majority of cases a proportion near to α will ϕ. So far, we have been considering *absolute* propensities of degree α to ϕ. But we may also consider *relative* propensities. Suppose some item can have one of properties $C_1, \ldots, C_n$. Then we can consider a relative propensity to ϕ, specified by the n-tuple of degrees $\langle\alpha_1, \ldots, \alpha_n\rangle$. This will be manifested by a collection of items each with this relative propensity, as follows:

> If the frequency of items with C_i is μ_i then, for a conforming collection, the frequency of items which ϕ will approximate $\nu = \alpha_1\mu_1 + \ldots + \alpha_n\mu_n$.

For example, we could take the C_i to be possible positions of a particle, and take ϕ-ing to be the coming to be in some specified position D a short time Δt later. In this case the frequencies will depend not merely on some intrinsic property of the particle, but also on the forces acting. However, if Δt is small, we may ignore the effect of the forces and assume that the frequencies depend only on intrinsic properties of the particles, which will, for that reason, now count as propensities. Since the C_i are different positions and D is the same position, we would expect the α_i to differ. We might for instance find that α_1 is high and the other α_i low. In that case the overall result is rather like a disposition to move to D provided the particle is at C_i.

At this point the reader may wonder what is the point of introducing these relative propensities. Would it not suffice to consider an absolute propensity to ϕ whose degree depended on which of the C_i the item had? This will not suffice if we complicate things further. In the above I said that a *conforming collection* is one for which the frequency of items which ϕ approximates $\alpha_1\mu_1 + \ldots + \alpha_n\mu_n$, where the μ_i are the frequencies of items with C_i. I shall now restrict my attention to a *sub-class* of the collections, namely, those for which the n-tuple of relative frequencies $\mu_1, \ldots, \mu_n$ with which $C_1, \ldots, C_n$ occur, belong to some specified non-empty set, *Priv*, the privileged set. (I assume of course, that for every n-tuple in *Priv*, $\mu_1 + \mu_2 + \ldots + \mu_n = 1$.) Let us now go back to clauses (iii) and (iv) of my hypothesis about propensities. Those were:

(iii) The proportion of large collections conforming to the propensities of their members is high.

and:

(iv) That the proportion of conforming collections is high is itself no accident.

Modify (iii) and (iv) so that we have instead:

(v) The proportion of conforming collections whose frequencies belong to *Priv* is high.

and:

(vi) That the proportion of conforming collections whose frequencies belong to *Priv* is high is no accident.

If (v) and (vi) replace (iii) and (iv) then we have what I call *tied* relative propensities. Tied relative propensities are manifested by collections for which the relative frequencies of the C_i are in the privileged class. The point of introducing relative propensities was precisely to make this further step to *tied* relative propensities.

Privileged classes arise naturally in quantum theory. For Heisenberg's Uncertainty Principle puts a constraint on the variances for position and momentum. Suppose we take the quantum state for a one-particle quantum system to represent a collection of particles each with a determinate position and momentum. Then we interpret this variance of position as the statistical spread of positions in the collection. In this case, then, we cannot prepare, or otherwise discover, a collection for which the product of the variances for position and momentum is less than indicated. Such collections exist, as subsets of the discoverable collections, but they are undiscoverable. It is not ad hoc to treat the discoverable collections as forming a privileged class. And a set of relative propensities could plausibly be tied to that class.

I shall develop this when I discuss the Statistical Interpretation in the next chapter. In anticipation I shall pull a rabbit out of my hat by showing how a set of tied relative propensities can take *negative* values. For example, suppose we consider only two characteristics C_1 and C_2. And let the privileged class consist of those pairs of frequencies each of which is between 1/4 and 3/4. Consider the tied relative propensity whose pair of degrees is $\langle -1/2, 3/2 \rangle$. If the frequencies for C_1 and C_2 are μ and $1 - \mu$, then ν, the frequency of ϕ-ing will approximate $3/2 - 2\mu$. As μ varies between 1/4 and 3/4, $3/2 - 2\mu$ varies between 0 and 1. So we are never threatened with negative frequencies. This example illustrates how negative propensities can arise, provided they are tied relative propensities.

V THE SEMI-KANTIAN SPECULATION

I have already rejected both phenomenalist and Kantian accounts of quantum theory on the grounds that if stated in phenomenal terms quantum theory loses all theoretical appeal. However, there is a semi-Kantian speculation with clear links to the Copenhagen School.[12] On it, a state is characterized by its propensity to produce certain results when measured. In accordance with the emphasis I have placed on

YNOs, I take the semi-Kantian speculation to characterize a state by its (absolute) propensity to give a Yes result to the various YNOs.

In this speculation, then, we are realists about quantum states, but they are described merely as propensities to produce Yes results for various YNOs. I call this a *semi*-Kantian speculation. A fully Kantian speculation is one in which we only know things by their (causal) relations to us. Now propensities are not relations, they are intrinsic properties. None the less to describe a property as a propensity is to describe it in terms of something else. So if the state is described in terms of propensities to affect observers then it is described in terms of the effect on others, which has a Kantian flavour to it.

Although I shall argue that the semi-Kantian speculation is mistaken, I concede that it is a natural one in the context of quantum theory. For suppose the probabilities mentioned in the discussion of von Neumann dynamics are interpreted as propensities. Then we shall be able to describe a quantum system in terms of its propensity to give Yes results to various YNOs made after varying periods of evolution in accordance with the Schroedinger dynamics. By appeal to the Radical Privilege, we might then insist that this is all there is to a quantum state.

As an illustration of something like this argument, consider some remarks of Fine's. He says: 'The point of the calculation is this. If probability presupposes chance and if chance need not attach to conjunctions whenever it attaches to the conjuncts, then neither does probability' (1973b, p. 12). Here Fine is summing up an argument whose conclusion is that we should not expect a joint distribution for position and momentum even given a standard theory of probability. Fine's argument is based on the following premiss: 'If ϕ is not a situation of chance for the display of a property P then we cannot correctly ascribe to any object in the situation ϕ any probability for P, other than 0 or 1' (p. 11). I take it that Fine is assuming that the probabilities involved in quantum theory are always propensities, and therefore closely connected to chances.

One criticism of the semi-Kantian interpretation is that even if we could describe all the properties of states as propensities, that would not make their description as occurrent properties redundant. In particular, consider a one-particle system in an eigenstate for the YNO of whether it is in region V. Although we could describe a certain property as having a propensity 1 to produce a Yes result for this YNO, we cannot avoid also describing it as simply being in region V.

(That this latter description is available if we need it follows from Weak Veridicality.) For the attempt to describe all properties as propensities leads us to an infinite regress: properties are described as propensities to bring about propensities etc. At no stage in the regress have we ever eliminated occurrent descriptions of properties. I conclude that although every property may well be a propensity, necessarily some of them must also be described occurrently. But the semi-Kantian speculation either denies this necessity, or at least *suggests* that there is no such necessity.

Another criticism of the semi-Kantian speculation concerns the propensities to produce Yes results for position YNOs. It is simpler (partly) to characterize a state in terms of the (perhaps indeterminate) position of a particle than in terms of a propensity to affect a measuring apparatus in a certain way. And I shall argue in two ways that we can (partially) characterize the state in this simpler fashion. I conclude that we should do so. Note that I am not criticizing *all* resort to propensities. For instance, on the Determinate Particle Speculation which I develop in the next chapter, the state is partly characterized by propensities for change of location.

My first argument starts from the premiss that there are eigenstates for the YNO of whether the particle is in region *V*. By Weak Veridicality such a state must be one in which the particle is in *V*. So sometimes particles have locations, albeit indeterminate ones. But in that case it seems unnecessarily complicated ever to treat the probability distribution for position as a propensity to produce Yes results for YNOs. Given that we *sometimes* have indeterminate locations, why should we not *always* construe what is apparently a probability distribution for location as an indeterminate location? (That we can so construe it is something I argue for in chapter 7.) Since we have indeterminate location anyway, it follows that we *should* so construe it. But in that case we no longer need to characterize the state using propensities to produce Yes results for location YNOs.

There is only one reply to this argument which I can think of. It is that there is a not too complicated way of distinguishing between cases in which a particle has location and cases in which it has not. We select a suitable class of *non-trivial*[13] position YNOs. Then we say that a particle has location just in case it is in an eigenstate for one of these YNOs.

My second argument shows this reply is not tenable. Suppose we grant that particles do not always have a location. Now consider an eigenstate for the YNO of having momentum in *W*, where *W* is a finite region. There is a dilemma. Either the particle has no location or it has one (although perhaps it is indeterminate).

In the first case, it is in the embarrassing position of having a momentum but no location. If momentum were defined as occurrent momentum this would be incoherent. Coherence can be retained either if momentum is analysed as a disposition to change location or if what we call momentum has been misdescribed and what is measured is in fact a disposition to change location. In either case the so-called momentum would be a disposition. And since the particle has no location, it would be an unmanifested disposition, like the brittleness of a vase which is never dropped. This establishes coherence. But coherence is not enough. For in this situation we would have the further peculiarity that it is physically impossible for the propensity to be manifested. For to manifest it, the particle would have to acquire a location. And that, we are assuming, would only happen in an eigenstate for a position YNO, which is not compatible with its continuance as an eigenstate for the momentum YNO.[14] What we would have posited then, would be a propensity or disposition which not only *will not* but *could not* be manifested. Furthermore, this impossibility is not just an impossibility for one kind of particle but an impossibility for any particle. Suppose, to take an analogy, no person ever displayed courage. None the less suppose it was posited that sometimes people were courageous, but that, as a result of human nature, exposure to danger resulted in loss of courage. This position is strictly coherent but not tenable. I conclude that particles must always have locations even if the location is sometimes indeterminate.

The other half of the dilemma is that some particles in eigenstates for momentum YNOs do have position. Now these eigenstates will be eigenstates for no non-trivial position YNO.[15] So if you continue to deny that all particles have (perhaps indeterminate) locations then you must distinguish the two cases of having location and not having location on some grounds other than being in an eigenstate for a (suitable) non-trivial position YNO. In this way, the objection to the first argument is removed.

The semi-Kantian speculation is not tenable. It should make way for a successor theory, still in the Copenhagen spirit, in which a particle has indeterminate locations, and in which the state is characterized by this, by spin, and by various propensities, such as the propensities governing the change of indeterminate location. I provide this successor theory in chapter 7.

VI AN APPRAISAL OF THE COPENHAGEN SCHOOL

Putting to one side the immensely valuable work done by the Copenhagen School in formulating quantum theory, I have concentrated on their metaphysical speculations. I remove the positivist taint, which I do not consider to be entrenched. In its place I propose the Radical Privilege as a way of overcoming the Conservative Presumption. And, related to this, I defend their right to conceptual innovation. If we need particles of indeterminate location then we are entitled to them (I shall rely on the same Radical Privilege when I help myself to negative propensities). I have also discussed the theory of propensities, which has its roots in the Copenhagen School. I reject complementarity as proposed by Bohr, although I postpone the question of whether both a particle and a wave speculation could be correct until chapter 8. Finally, I argued that a semi-Kantian speculation in the spirit of the Copenhagen School should make way for a successor theory, that in which particles have indeterminate locations.

NOTES

1 We could also include Dirac and von Neumann.

2 It is possible to be a positivist of sorts and still accept inference to the best explanation, and hence the existence of unobservables. For a version of Positivism noted by Ayer (1971, p. 12) allows statements to be meaningful even though they are only verified in the weak sense of rendered probable. Hence Positivism does not commit one to Instrumentalism. However, this is indeed a weak version of Positivism. Historically positivists have been instrumentalists.

3 To be *empirical* is to have the intellectual virtue of making general theories (one's own or others') the servants rather than the masters of particular judgements based on experience. It does not imply being an *empiricist*. An empiricist is someone who holds a specific, to my mind narrow, theory of what constitutes experience, namely that only the having of sensations and the making of introspections count as experience.

4 For simplicity I consider the one-dimensional case. And I shall follow Messiah (1961, p. 131). If the state is represented by a function of location:

$$\Psi(q) = \exp [ip_0 q - (q - q_0)^2/2\xi^2]$$

then its 'momentum-space' representation is:

$$\Phi(p) = \xi \exp [iq_0(p_0 - p) - 1/2\, \xi^2(p - p_0)^2]$$

The smaller ξ the smaller the variance for q and the larger the variance for p. This establishes a continuum between the small ξ case where the position 'picture' applies, and the large ξ case where the momentum 'picture' applies.

5 I mention the Feyerabend–Hooker argument only as an *example* of how a consideration apparently supporting Bohrian Complementarity can be deflected, so that, if it supports anything, it supports the Indeterminate Particle Speculation quite as much as Complementarity. I do not, in fact, accept it. It is based on the way that in some systems (e.g. an electron in a hydrogen atom) every quantum state is an eigenstate for the YNO of whether the energy is in a specific set of values E which contains a series of discrete levels. Thus e_0 the lowest energy, and e_1 are in E but no point between them is. This raises problems concerning what happens when the energy level changes as a result of interaction with some other system, such as the electromagnetic field. It is argued that a discontinuity in determinate energy is not plausible. Hence it is concluded that at some time there is no determinate energy. I reject this. For an alternative conclusion is that the energy of the composite system (electron plus electromagnetic field) is not analysable as the sum of energies of the two component systems (because the potential energy due to the forces of interaction is part of the energy neither of one system nor of the other). As a consequence, the quantum state of the electron is not an eigenstate for the YNO of whether the energy is in E, in the cases when the electron is interacting with the electromagnetic field. (Actually it is *always* interacting with the electromagnetic field but for some purposes we can ignore this and consider the electron in isolation.)

6 It is tempting at this point to consider infinite, even non-denumerable, collections. These are often called *ensembles*. It might then be claimed that in an ensemble of systems with propensity of degree α to ϕ, a proportion exactly equal to α will ϕ. There are two reasons why we should resist the temptation to consider infinite ensembles. The first is that we have a surer intuitive grasp of large finite collections than of ensembles. The second is that the same difficulty arises, even in an infinite collection. For by chance, only a small proportion of systems might ϕ. The probability

of this happening will be infinitesimal. But that does not imply that it cannot happen.

7 This identification of the disposition with an occurrent property (called its *categorical base*) is not trouble-free. For there is the well-known difficulty that the same disposition can be based on *different* categorical bases (Mackie, 1973, p. 148). One response to this is to identify the disposition with a disjunctive property. Another is to deny that there is strictly speaking a property of, say, *brittleness*, and to say an object is brittle just in case it has *some* property which is such that it breaks when dropped in normal circumstances. (For a discussion of these questions, see Prior, 1985, ch. 6.) As a consequence of these difficulties, my account of propensities might fail as a piece of *conceptual analysis*. But my present purpose is not conceptual analysis. It is to speculate about the property something *actually has*, when it has a certain propensity.

8 Mellor (1971, pp. 70–6) distinguishes propensities from chances. On his account, a propensity is a disposition which displays itself, in a given situation, by means of chances. There are two aspects to this distinction. The first is that while chances are properties of the total set-up, propensities are properties of some item which is just part of the set-up. In a quantum-theoretic context, a propensity to give a Yes result to some specified YNO is a property of a quantum system. The chances are properties of the set-up which includes the measuring apparatus. Again, a propensity to change location would be a property of a particle, while the chance of a given change is a property of the set-up which includes the forces acting on the particle. In this first respect I follow Mellor, treating the propensities of various items as indeed intrinsic properties of those items. However, he also insists that propensities are dispositions displayed by chances. I see no need for this two-stage approach. I treat propensities much as Mellor treats chances, namely as something knowledge of which leads to suitable degrees of belief. The degree of the propensity will equal the chance *in a standard set-up*. For measuring a YNO the standard set-up is that provided by an idealized measurement for which von Neumann dynamics holds. For change of location a standard set-up is a situation where forces are neglible.

9 I do not assume that there is anything *intrinsically* numerical about degrees of belief. For a discussion of their numerical representation see Forrest (1986, ch. 4).

10 To obtain proportions we need to consider only finitely many systems. If there are an infinity of particles in the universe, which I doubt, then we would have to localize this account. We could for instance insist that (iii) holds for any large enough region of the universe.

11 The Dretske–Tooley–Armstrong theory that laws of nature are relations between universals is developed by Armstrong (Armstrong, 1983).

Armstrong considers (pp. 135–6) laws which ensure that a certain proportion of *F*s are *G*s. What we might require for propensities on my account is such a law where *being an F* is *being a collection of a certain kind*, and *being a G* is *being a conforming collection*.

12 This speculation is, I suspect, part of the motivation for the fashionable logico-algebraic approach to quantum theory, discussed in the last sections of chapter 2.

13 The trivial position YNO is whether the particle is somewhere. The answer is Yes with probability 1, for any state.

14 No state is an eigenstate for both a (finite) location YNO and a (finite) momentum YNO (see ch. 2, note 33).

15 See note 14.

5

The Determinate Particle Speculation

There are two rather different speculations which require realism about particles. One, which I call the Indeterminate Particle Speculation, allows that the particles being considered may lack a sharp or determinate location. The other, which I discuss in this chapter, insists that every particle has a determinate location at all times. For simplicity I shall suppose a location is completely specified by a point, either because the particle is itself a point particle or because we are to consider its centre of mass.

This sort of position has been defended by Landé (1965), Popper (1982, pp. 125–43), Audi (1973) and Fine (1973a,b). Like them I shall also assume that the particles have determinate momentum, *provided* that is simply defined as the occurrent momentum, namely mass times velocity (in the non-relativistic case). But if momentum is considered as a disposition to change location, then I shall deny they have momentum, replacing it by a suitable set of propensities to change location. The resulting speculation can thus be thought of as an elaboration of Popper's conjecture, which involves determinate particles and propensities.

On the Determinate Particle Speculation there are truths which could not be discovered. For, by Heisenberg's Uncertainty Principle, we cannot discover both the determinate position and the determinate momentum of a particle. Yet (on the Determinate Particle Speculation) there is some truth about the determinate position and determinate momentum of the particle. This incapacity, even in principle, to discover all truths should not surprise us. Why should we ever have thought otherwise? However, the proper pride in the achievements of scientists has often driven out an equally proper humility. As a result, this, quite innocent, feature of the Determinate Particle Speculation could be considered a defect. But it should not be. Hidden truths are no scandal.

There are, however, a number of more serious objections to the Determinate Particle Speculation. My aim is to provide a version of this speculation which meets those objections. For the most part my strategy will be to examine what might initially look like an unacceptable consequence of the Determinate Particle Speculation, and to argue that this consequence is, on reflection, quite acceptable.

I NO HIDDEN VARIABLES?

A quantum state does not specify even an approximately determinate location for a particle unless it is an eigenstate for a suitable position YNO. Therefore, if we none the less speculate that particles do have a determinate location at all times, we are positing *hidden states*, which, unlike the quantum states, do specify a determinate position at all times. On my development of this speculation, the quantum state describes a large collection of systems each in some hidden state or other. But I shall delay until the next section the justification for this. Here, as a preliminary, I shall discuss the famous No Hidden Variables Theorems.

I am not proposing a rival hidden variables theory to quantum theory. My task is to speculate about quantum theory itself. But this leads to a problem. Are there not various results due to von Neumann (1955), Jauch and Piron (1963), Cohen (1966), and Kochen and Specker (1967), which show that I am attempting the impossible?

I concede that these results show that we cannot treat quantum theory just like classical statistical mechanics. But they do not show that we cannot have a determinate particle speculation of the sort I am proposing. I shall concentrate on the results of Cohen (1966) and of Kochen and Specker (1967). My reason for ignoring the original work of von Neumann is that the Kochen and Specker result (1967, pp. 81–2) is the more powerful. I ignore the result of Jauch and Piron because it is more powerful than von Neumann's only in the sense that it is more general, not requiring all the usual formalism. And my aim is to provide a 'hidden variables' speculation even given the standard formalism.

First I consider Cohen's result (1966). In one respect this is less general than that of Kochen and Specker; but it has the advantage of being more accessible. Moreover, the Determinate Particle Speculation which I propose is one to which Cohen's result is applicable. For I consider the state (of a spinless particle) to be given by a 'joint distribution for position and momentum'. Now it so happens that I consider this to be something of a misdescription. But that does not prevent Cohen's result from being applicable.

Consider a one-particle quantum system and ignore spin. The sort of hidden state speculation under discussion is that a quantum state (even a pure one) does not describe a single system, but rather gives us a probability distribution for the hidden states of systems. The hidden state is one in which the particle has a determinate position and a determinate momentum. It will turn out that these probability distributions can somehow take negative values. This is *one* departure from classical statistical mechanics. We might hope it is the *only* one. But Cohen's result shows it is not. For in classical statistical mechanics we have *Universal Statistical Veridicality*, namely Statistical Veridicality for all observables.

Cohen's result shows that if we propose this kind of 'hidden variable' speculation then we must abandon Universal Statistical Veridicality. Indeed he shows that we cannot have statistical veridicality for all four of the following:[1]

(i) Position.
(ii) Momentum.
(iii) A suitable function of position and momentum (e.g. a component of angular momentum).
(iv) A suitable function of the observable (iii) (e.g. the square of the component of angular momentum).

Cohen's result shows that we should abandon Universal Statistical Veridicality when proposing a determinate particle speculation, a conclusion reinforced by the more general result of Kochen and Specker (1967), discussed in detail by Redhead (1987, ch. 5). This result concerns *any* hidden state speculation, whereas Cohen was restricting his attention to ones in which the quantum state gives a probability distribution over position and momentum.

Like the other No Hidden Variables Theorems that of Kochen and Specker requires several premisses which may be questioned. One of these is Universal Statistical Veridicality.[2] Given these premisses, they obtain the result that a hidden variable speculation is only tenable for the very special case of spin (for spin one-half particles). Indeed they show precisely which probability distribution is required if all spin YNOs are to be statistically veridical (Kochen and Specker, 1967, pp. 75–80). This distribution is not the same as the one I argued for when discussing the problems of mixture in chapter 3. So there may be some reason for denying statistical veridicality for spin YNOs in spite of their exemption from Kochen and Specker's result.

I have already denied Universal Statistical Veridicality (see chapter 3, section III), and the result of Kochen and Specker poses no further threat. But it is worth commenting on the other premisses required for that result. Here I rely on the detailed analysis of Redhead (1987, ch. 5). He considers three theses, which he calls *View A*, the *Correspondence Rule* and the *Reality Principle*. Assuming only the formalism[3] and a quite minimal interpretation,[4] Redhead (p. 45) proves that at least one of the three theses is false.

View A is the thesis that given any observable and any (hidden) state of the system then the observable has a sharp (i.e. determinate) but perhaps unknown value. This is equivalent to saying that for any YNO and any hidden state there is either a Yes or a No result. So it is a bivalentist position.

Quite how strong View A is depends on what is said to be an observable. The Correspondence Rule (p. 133) tells us that observables are in one-to-one correspondence with the appropriate operators in the formalism. On the one hand, View A excludes the case where some of these operators are not treated as genuine observables. On the other hand, it excludes the suggestion that the same operator represents more than one observable.

If we assume the Correspondence Rule then the Reality Principle (p. 137) implies that the observable represented by a given operator is measured by the statistical distribution of results which the formalism associates with the operator. Hence, the Reality Principle amounts to the assumption of Universal Statistical Veridicality.

Redhead's analysis, then, raises the question of why I have, like Fine, rejected Redhead's Reality Principle, rather than View A or the Correspondence Rule. View A is just a statement of the Determinate Particle Speculation, in which it is assumed that all observables have determinate values.[5] There is, perhaps, room for a speculation in which some do and others do not. But in this chapter I am defending the tenability of a thorough-going Determinate Particle Speculation. So View A is not in dispute.

My reason for not abandoning the Correspondence Rule, as van Fraassen does (1973), is the obvious one. Redhead tells us that van Fraassen's proposal was referred to by Clark Glymour as *de-Ockhamizing* quantum mechanics (1987, p. 137). And this is precisely the objection. I would be the first to concede that Ockhamist considerations establish no more than a presumption. In this case the presumption might be overcome if Universal Statistical Veridicality were a desideratum for a Determinate Particle Speculation. But as I argued in chapter 3, it is not.

It is worth comparing my response to the No Hidden Variables Theorem with that of Fine (1973b). Fine argues that we do not need a joint probability distribution for position and momentum even on a Determinate Particle Speculation. Initially, this would seem quite incompatible with my own response, which is to retain a joint distribution, but to reject Universal Statistical Veridicality. Actually there is rather less disagreement here than might appear at first sight. Fine is considering probabilities based on a primitive notion of the *chance* of the display of a property. On the one hand, this commits him to a fairly strong connection between the probability of particles having some specified property and the frequency of observations of that property. For this reason, he should be reluctant to abandon Statistical Veridicality. On the other hand, he should not be expected to believe in a joint probability distribution unless position and momentum were jointly observable, which they are not. By contrast, I am working with a more general concept of probability, which allows me to propose a joint distribution, but which removes the pressure for Universal Statistical Veridicality.

I have concluded that the result of Kochen and Specker is no threat to a Determinate Particle Speculation. A possible objection here would be that further analysis of the Kochen and Specker result might show that they do not require *Universal* Statistical Veridicality, but merely Statistical Veridicality for some specified observables.

In reply to this it is worth commenting on the observables being considered in the proof of Kochen and Specker's result. The crucial assumption is that we have Statistical Veridicality not merely for some observables such as position and momentum but for any *function* of them (Kochen and Specker, 1967, p. 63; Redhead, 1987, p. 121). To take a representative example, we require Statistical Veridicality not merely for some observable A, but also A^2. Let us suppose that the errors in measuring A tend to balance out and so Statistical Veridicality holds for A (see chapter 3, section III). Then it should be *expected* to fail for A^2. As in my earlier discussion, this is most clearly seen for small values. Suppose the value of A is very close to zero, or even zero itself. Then we can have errors either way in the measurement of A which balance out, since A could take negative values. But for A^2 the errors cannot cancel out in this way.

More generally, there is an argument to show that we should expect failure of Statistical Veridicality for any non-linear function of A, that is, for any function not of the simple form $cA + d$, where c and d are constants.[6] Thus there is no threat from a more detailed analysis of the results of Kochen and Specker.

II VARIANTS OF THE DETERMINATE PARTICLE SPECULATION

Associated with a given quantum state are various probabilities for Yes result for position YNOs. What connection have these probabilities with the particle, which, we are assuming, has a determinate point-location? By Weak Veridicality, in an eigenstate for a Yes result for the YNO of whether a particle is in a region V, the particle must indeed be somewhere in V. However, within this constraint there are three variants of the Determinate Particle Speculation, based on three different sorts of probability which we might be considering. The probabilities could either be subjective probabilities, propensities or approximate frequencies. Later I shall discuss the advantages and disadvantages of these three interpretations, but in this section I shall merely expound them.

I have already mentioned subjective probabilities. To recapitulate, by the *subjective probability* that p, I mean a number from 0 to 1 which *represents* a person's degree of belief[7] that p. I do not assume that degrees of belief are intrinsically numerical quantities. But provided we are considering degrees of belief satisfying various consistency constraints we may *represent* them numerically. To cut a long story short:[8] if there is the same degree of belief in every member of a set of n pairwise inconsistent propositions whose disjunction is believed with certainty, then we represent that degree of belief by 1/n. And that puts constraints on the representation of other degrees. I shall assume, then, that there are subjective degrees of belief, at least for the sorts of propositions being considered in quantum theory. The consistency constraints are fairly weak, the following being typical:

> Suppose p is inconsistent with q, and r is inconsistent with s. Suppose p is believed with greater confidence than r. And suppose q is believed with greater or equal confidence than s. Then (p or q) is believed with greater confidence than (r or s).

Such constraints are compatible with a great deal of irrationality in what you believe. For the constraints do not ensure that a person reasons inductively, that she rejects ad hoc hypotheses and so on. Because of the weakness of the constraints presupposed when degrees of belief are represented, a quantum state u will not always specify the actual degree of belief someone has that a given position YNO will have a Yes result, even if we assume the person understands and accepts quantum theory. For people can and often do have a quite unwarranted confidence in theses they have fallen in love with, without violating any consistency constraints. So the state u would have to *prescribe* rather than *describe* the degree of belief. The probability assigned to a position YNO would represent the only[9] *warranted* degree of belief given the evidence available. This introduction of the warrantedness qualification removes the *subjectivity* of the probabilities in one sense of that abused word. However 'subjective probability' remains a convenient label for this interpretation of probability.

I have already discussed propensities. On the propensity interpretation, which is *suggested* by Popper's position,[10] the state is partly characterized by the absolute propensities for position YNOs to have Yes results.

On the frequency interpretation, the probability for region V is to be interpreted as the approximate[11] frequency of particles in V in a large collection. So if the theory predicts that the probability of a Yes result for the YNO of being in V is some value x, then it is taken to be predicting that the proportion of systems for which the particle is in V approximates x. Now the quantum measurement theory tells us that the frequency of measurements recording that a particle is in region V will also approximate this probability. Hence on the frequency interpretation we have not merely Weak Veridicality, but also Statistical Veridicality for location.

It might be objected that any conceptual analysis of probabilities as (approximate) relative frequencies must fail because, by sheer luck, any probability other than 0 or 1 could co-exist with any frequency, so we might not have the required approximation. This point about conceptual analysis is one I accept. So a frequency interpretation of probability is an *eliminative* or *replacement* thesis, rather than an *analysis*. On it we propose replacing probabilities by frequencies. Hence it would be more accurate to talk of the '*probabilities*' rather than the *probabilities*. I hope I do not have to labour the point that such a replacement is quite legitimate as a speculation. This follows from the Radical Privilege, for we have no evidence for probabilities except via the frequencies.

I conclude this section by noting two connections between the variants. The first is between frequencies and warranted degrees of belief. Suppose we have a large collection of particles. Suppose we know that a proportion x of members will give a Yes result for some position YNO. And suppose we are considering one member of the collection. Further suppose that we have no information relevant to the prediction other than that the system considered is a member of the collection. Then, given the way we represent degrees of belief, the only warranted degree of belief that there will be a Yes result is the degree x (see Forrest, 1986, ch. 8).

Secondly, there is a connection between propensities and warranted degrees of belief. Consider someone who accepts quantum theory, the metaphysical speculation here being considered, and various pieces of evidence, and who therefore comes to be certain that the propensity for a Yes result for a given position YNO was x, and who has no further relevant evidence. (In particular I assume she has not observed the result.) If she has a degree of belief other than x that there will be a Yes result, then she is irrational. Hence the greater the propensity the greater the warranted degree of belief.

Again, suppose we have a collection of systems. Consider a rational person who believes that a proportion x of systems in the collection are such as to give Yes results, and who has no further relevant evidence. She must come to have a degree of belief represented by x that some specified member of the collection will give a Yes result.

These connections show that, far from the three variants of the Determinate Particle Speculation being incompatible, if any of them holds then the Subjective Probability variant also holds. But we may seek to explain one variant in terms of another. Accordingly, in the next section I shall argue that *subjective probability cannot stand alone*:

> Unless the person whose degrees of belief are being considered already has beliefs about either propensities to give Yes results to YNOs, or about frequencies of particles with locations in various

regions, then a degree of belief as specified by the Subjective Probability Variant would not be warranted.

In section IV below I argue that quantum systems do *not* have propensities to give Yes results to location YNOs. This leaves a variant of the Determinate Particle Speculation in which the probability distribution for position is primarily treated as a frequency distribution for a large collection (of actual or merely possible systems), but, as a result, provides a constraint on the warranted degrees of belief of someone who knows of this distribution of frequencies. It should be noted, though, that I shall combine this frequency interpretation of the probability distribution for location with a propensity interpretation for the probabilities for change of location. To anticipate, this is because we must then consider negative probabilities which are most easily understood as negative propensities. This combination of frequencies for location with propensities for change of location is a harmonious one. It makes no obvious sense to talk of a propensity to *be*, but plenty of sense to talk of a propensity to *become*.

III SUBJECTIVE PROBABILITY CANNOT STAND ALONE

Consider a one-particle system. Suppose we have, quite rationally,[12] various subjective probabilities that the system would give a Yes result to various YNOs at some specified time. Then, using quantum theory, we can calculate the probability distribution for Yes results for YNOs at a later time. Now, like any scientific theory, this has implications for which degrees of belief are warranted. We should indeed have degrees of belief which are in agreement with the probabilities which have been calculated. I do not deny that. What I do deny is that this constraint on the degrees of belief could be imposed by quantum theory if the probabilities were themselves nothing more than warranted degrees of belief. For there to be a constraint on degrees of belief, the probabilities must be something more. Otherwise quantum theory is no longer a descriptive theory about certain recondite (non-evaluative) facts. It would itself become an evaluation of how we ought to reason, or of the degrees of belief we should have. As such it would be a piece of Logic, or perhaps Epistemology, not a scientific theory. But quite clearly quantum theory is a description not an evaluation. I grant that *some* revision to logic or probability-theory, and hence to the constraints on our degrees of belief, is a prima facie admissible speculation. (Although I have argued against this.) But denying distributivity is one thing, insisting that Schroedinger's equation is a piece of logic is another: quantum theory is primarily descriptive.

I now provide a connected but more specific criticism. First I claim that the constraint on the subjective probabilities holds only for someone who has accepted quantum theory with complete confidence. This claim is argued for as follows:

> It is not irrational to have some doubts about quantum theory. Hence the constraint provided by the observations alone must permit a wider range of degrees of belief than those indicated on the Subjective Probability Variant being considered. To make the subjective probabilities agree with the predicted probability we must therefore consider the degrees of belief of someone who is stipulated to have accepted quantum theory with total confidence.

On the proposed variant, then, quantum theory specifies various probabilities which are themselves interpreted as the degrees of belief which a rational person would have who, among other things, accepted quantum theory itself. This circularity makes the variant quite trivial. It is simply saying that someone who accepts some theory T (in this instance quantum theory), should have degrees of belief as constrained by T. And *that* would hold even if T were a complete mistake. To avoid this trivialization something other than the subjective probabilities must be constraining the degrees of belief. The only candidates for what else could thus constrain the degrees of belief are the probabilities interpreted in some other way. Hence the Subjective Probability Interpretation cannot stand on its own.

IV LIMITING THE ROLE OF PROPENSITIES

I shall argue below that a quantum state does indeed describe propensities, namely tied relative propensities for change of location. However, I deny that quantum systems have propensities for giving Yes results for various location YNOs. For such a propensity would be an intrinsic *property* of the system, whereas I claim that the probability for the Yes result for a location YNO is the proportion in a collection of systems of those for which the Yes result is veridical.

My criticism of the thesis that there are propensities for Yes results for location YNOs is that this ought not to be combined with the Determinate Particle Speculation. Otherwise there is an objection based on Schroedinger's Cat. (A similar objection arises with the twin-slit experiment.) Consider a state corresponding to a 50 per cent probability of an intact and a 50 per cent probability of an exploded cat. We have a Determinate Particle Speculation, and being exploded is a matter of the location of all the parts of the cat. Hence the cat either is or is not exploded. Say it is exploded. On the proposed variant it would have a propensity to become unexploded on being observed. A wonderful cat indeed! To avoid this difficulty the propensity interpretation would have to be combined with some way of ensuring that the von Neumann dynamics occurred earlier, at the intended time of the explosion. In that case the propensity is no longer a propensity to give a Yes result to observations, but a propensity to change location in such circumstances, which would have to be specified. So this interpretation collapses into the one I shall advocate, in which the propensities involved are propensities to change location rather than propensities to give Yes results when making observations. My aim is not exegesis, but, for what it is worth, I interpret Popper as considering propensities to change location rather than propensities to give Yes results to YNOs.

The only reply to this objection which I can think of appeals to the hypothesis that not only the present and future can be affected by an observation but the past also. Using that hypothesis we could argue that the cat's being observed at, say, 6 a.m., to be exploded, causes its having been exploded from 2 a.m. This avoids a propensity to become intact again after being exploded. Now I accept backward causation as a legitimate speculation, but appealing to it in this case is to include a quite unnecessary and ad hoc hypothesis. Because we can interpret the state as describing the frequency of locations in a large collection, we do not need to resort to backward causation.

V TWO PROBLEMS FOR THE FREQUENCY INTERPRETATION

As a result of the discussion of the previous two sections, I shall assume that the Determinate Particle Speculation requires a frequency interpretation of the probability-distribution for location.

Initially we thought of a pure state as a state of a single system. On a frequency interpretation this is not the case. All quantum states, even the pure ones, are now considered to be states of large collections of systems rather than of individual systems. First, consider position. The probability distribution for Yes results for the position YNOs is treated as a distribution of frequencies for position in a collection of systems. The prediction about such frequencies will be more accurate the larger the collection. Now consider momentum. The obvious frequency interpretation is to treat the probability distribution for momentum just like that for position. It might be proposed that individual particles have determinate positions and momenta, and that the state describes the separate distributions for position and for momentum in some collection.

The first problem is that the distribution for position and the distribution for momentum do not jointly specify the state, even if the particle is spinless. Some further account has to be given of the information contained in the state. One proposal is that the state is specified by a *joint* distribution for position and momentum. Hence it would describe the frequencies with which particles of various positions and various momenta occur in the collection. As I mentioned in chapter 2 (section VI), some of these frequencies would have to be negative. Hence we would have to make sense of negative frequencies. The best strategy here would be to reinterpret apparent frequencies of, say, electrons as the difference between the frequency of positive-mass electrons and that of negative-mass electrons. Negative-mass electrons would be like electrons except they have negative mass.[13] I shall discuss something very like this proposal in chapter 7. However, in response to the *next* problem I shall provide an alternative interpretation of the joint distribution, one which will involve negative propensities for change of location. Hence we do not require negative-mass particles on the Determinate Particle Speculation.

The second problem for a frequency interpretation is that we often do not have a large collection but only an individual system. However, we still assume the theory puts a constraint on the system's dynamics. One proposed solution (the Many Worlds Theory) to this problem is to propose a large (perhaps infinite) collection of sub-universes.[14] In each one there will be a copy of what we usually think of as the whole universe and hence of the system being considered. This ensures that we always have a large collection. If necessary I would posit these sub-universes. But it is not necessary. For there is no need to consider an *actual* collection at all. To consider the constraints put on a single system, it is enough to consider what would happen if we *did* have a suitable collection. What we do is this: we describe various possible systems in such a way that the description does not tell us which of the systems is actual.[15] And we select the frequencies in this collection of possible systems so as to approximate the subjective probabilities. Then the subjective probabilities are conveniently summed up by saying that there is (approximately) the same degree of belief that any one possible system is actual as that any other one is. Now we consider the evolution of the collection of possible systems in accordance with the Schroedinger dynamics. We still have (approximately) equal degrees of belief that the actual system is any one of the members of the collection. And this then constrains degrees of belief about the state of the actual system, just as in the Subjective Probability Variant. By considering possible systems, therefore, we can handle the case of a single actual system.

Clearly this solution is more economical than the sub-universe (Many Worlds) solution. But it leads to a dilemma, which I call the Redundancy/Interaction Dilemma. One horn of this dilemma is that the frequency interpretation is threatened with redundancy. If the collection can be a merely possible one, why do we need to consider it at all? For it is implausible that the merely possible affects the actual. A similar problem holds on the Many Worlds Theory. It is implausible that the behaviour of a system in one world should affect that in another. So, once again, why do we need a statistical interpretation at all? Indeed this horn of the dilemma is quite general. The Schroedinger dynamics preserves mixtures of states.[16] That is, if we have a number of sub-collections and mix them, we get the same result by mixing first and then letting the collection evolve with time as we do if we consider the sub-collections to evolve independently and then mix them. This supports the thesis that there is no interaction between members of the collection.

The other horn of the dilemma is that the probability distribution for location is *essentially* involved in the Schroedinger dynamics. To show this, I rely on Wigner's distribution. For although Wigner's is not the only possible formula, it is the simplest (Cohen, 1966, p. 319). Using the formula, it can be shown that the probability distribution for momentum *for a given location* changes in a way which depends on the probability distribution for location.[17] Thus the probability distribution for location is essentially involved in the Schroedinger dynamics.

I grasp the first horn of the dilemma, namely that there is no interaction between members of the collection. In particular, the actual system is not influenced by what is merely possible. But I resist the threat of redundancy. I do so by proposing that (for pure states) the frequency distribution for a collection of (possible) systems enters into the description of the actual system under consideration. It does so because the hidden state of the system is given not merely by the location, but also by the propensities for change of location. And my hypothesis is that these are *tied* propensities which (directly) manifest themselves only for a single privileged distribution of frequencies for location. The frequency distribution, and hence the collection of possible systems, enters into the description even of the hidden state, but only because it specifies the privileged class.

For the *hidden* state, then, the frequencies enter into the description only via the tied propensities. However, the quantum state, on this speculation, stands to these hidden states much as the states of statistical mechanics stand to those of classical mechanics. A *quantum* state indeed describes a large collection of systems, either actual ones or possible ones. Hence the probability distribution for location is involved in a pure quantum state in two ways. First it specifies the privileged class for the tied propensities of the members of a collection described by the pure state. Second, it specifies the frequency distribution for the location of the members of that collection. That the two frequency distributions coincide amounts to the assumption that pure states describe *standard propensity pure* collections, as I shall call them in the next section. This assumption does not apply to *mixed* states. Indeed, it could be used to explain the significance of the pure/mixed distinction.

I have argued that, for the case of a single particle, even a pure quantum state describes a large collection of particles. It specifies *both* the frequency distribution for the location of particles *and* various properties which are describable only as tied relative propensities for change of location. Conversely, these jointly specify the quantum state.

If we suppose that the particles move smoothly enough for them to have occurrent momenta, there will be a joint frequency distribution for position and occurrent momentum. However, I do not believe that the quantum state specifies in any way this joint distribution. Admittedly if the propensities for change of location had not been tied, then they would have been manifested by a frequency distribution for occurrent momentum for a given location. Hence the state would have specified a joint distribution for position and occurrent momentum. Moreover, we would have been under considerable pressure to adopt Wigner's formula for that joint distribution, with the negative values. However, as it is, I have posited tied relative propensities for change of location. Hence, they are not manifested by collections of particles all with the same location. Therefore they do not constrain the frequency distribution for occurrent momenta for particles *of a given location*.

VI A SINGLE DETERMINATE PARTICLE

I shall now describe in greater detail the case of a single particle, beginning with the case where there is no spin. I rely on Wigner's formula (Wigner, 1971).[18] This gives us what is *ostensibly* a joint distribution for position and momentum. I shall re-interpret this in accordance with the conclusions I reached in the previous section. For each region V of space and set W of momenta, Wigner's formula assigns a (possibly negative) probability $\Pr(V, W)$. The first step in the reinterpretation is to analyse $\Pr(V, W)$ in terms of the probabilities for position alone, and what are ostensibly the relative probabilities for momentum given position. The probabilities for position alone are given by the assignment of $\Pr(V)$ to every region V, where $\Pr(V)$ is obtained from $\Pr(V, W)$ by putting W as the set of all momenta. The relative probabilities are given by the assignment of $\Pr_q(W)$ to each location q and each set W of momenta, where if V is a small region around q, $\Pr(V, W)$ is approximately $\Pr(V) \times \Pr_q(W)$.

In accordance with the frequency interpretation of the probabilities for location, I take the quantum state, whether mixed or pure, to describe (approximately) a large collection of possible one-particle systems. The frequency with which the particle is in V is then (approximately) $\Pr(V)$.

In place of a disposition to move I posit that a single particle has a set of tied relative propensities for change of location. For each location q and each set W of momenta, I posit a (perhaps negative) propensity $\text{Prop}_q(W)$ to change location in such a way that the occurrent momentum is in W. This is the limit (as the time Δt tends to zero) of the chance of a change of location from q to some point q^*, such that the displacement from q to q^* is a vector Δq, where

$m\Delta q/\Delta t$, the average momentum over the time interval, is in W. Should we identify the $\mathrm{Prop}_q(W)$ with $\mathrm{Pr}_q(W)$? Not necessarily. For perhaps different members of the collection have different sets of relative propensities. In that case $\mathrm{Pr}_q(W)$ will itself be a mixture of $\mathrm{Prop}^1{}_q(W)$, $\mathrm{Prop}^2{}_q(W)$ etc. for the various different sets of propensities which occur. Since propensities are manifested statistically, we could not distinguish such a mixture from a collection in which all the particles have the same sets of propensities.

I shall call a collection in which all the particles do have the same (relative) propensities for change of location, a *propensity-pure* collection. I shall call a propensity-pure collection in which the distribution of locations is a member of the privileged class for that propensity a *standard* propensity-pure collection. For simplicity of exposition I shall assume that every standard propensity-pure collection is described by a pure quantum state and every pure quantum state provides such a description. (However, I leave open the possibility that some mixtures of standard propensity-pure collections can also be described by pure quantum states.) In this fashion, then, a quantum state for a spinless particle describes both a frequency distribution for position and the propensities for change of location.

There is a variant of the Determinate Particle Speculation worth mentioning. Instead of positing tied relative propensities $\mathrm{Prop}_q(W)$ for each location q, and each set of momenta, we could posit only the *propensity current*. If we do posit the tied propensities then the strength of the propensity current at q is defined as the mean or expected velocity at q given by the $\mathrm{Prop}_q(W)$. (Using Wigner's distribution it is easy to derive a formula for this current.[19]) But if we do not posit the tied propensities we could think of the current itself as a property that governs the change of location.

Under a fairly weak assumption, the propensity current and the frequency distribution for location jointly determine the quantum state.[20] Hence we might appeal to the Radical Privilege to deny that there are any tied relative propensities corresponding to the $\mathrm{Prop}_q(W)$, and to rely instead on the propensity current.

In this case, we would have a new replacement for dispositional momentum, namely the propensity current. It would specify the tendency for a collection of systems to evolve in various ways. The prima facie advantage of this is that we no longer have to consider negative propensities. However, I claim that this is not a genuine advantage, since we have given an adequate account of negative propensities. Furthermore, the tied propensities have the advantage of directly governing the motion of individual particles, whereas the current directly governs only the evolution of the whole collection. Thus the propensity current variant does not handle the Redundancy/Interaction Dilemma as well as the speculation involving tied propensities. For that reason I reject it.

For particles with spin we may provide a suitable generalization of Wigner's formula.[21] The interesting aspects of this can all be illustrated by ignoring both location and momentum and concentrating on the spin itself. Then for every pure state u there will be a direction a such that u is the eigenstate for the YNO of whether the spin is 'up' with respect to a.[22] As I indicated in chapter 3, we do best to posit a probability distribution uniform for those directions making an acute angle with a. This explains how a mixture of 50 per cent spin-up and 50 per cent spin-down gives the same quantum state regardless of the up direction.

Should we take the probabilities for spin to be relative frequencies (as for location) or propensities (as for momentum)? I consider spin to be intrinsic angular momentum. So, on the one hand, there is reason to treat it like momentum. That is, the probability of a given spin should be treated as a propensity for rotation. For momentum we required negative propensities. But we have no negative probabilities for spin. So this complication does not arise. The propensities would be relative to the particles' location, but they would not be tied to privileged frequencies for location. (This reflects the possibility of jointly observing both spin and location.)

On the other hand, if we take the probability for spin as a propensity for rotation we have to solve, somehow or other, the problem that these propensities will be manifested by rotations giving a peculiarly high occurrent angular momentum.[23]

Let me stress that it does not matter very much which way we interpret the probability for spin. And we may not be able to decide.

Finally I consider whether any veridicality theses hold other than Weak Veridicality. First, the probability $\Pr(V)$ is, according to the formalism of quantum theory, manifested in the frequencies for observation of position in V. Therefore location YNOs must be statistically veridical. That is, the frequency of Yes results for the position YNO of whether the particle is in V must approximate the frequencies with which the particles are actually in V, with the approximation getting better the larger the collection. We also have Statistical Veridicality for occurrent momentum.[24]

If the collection of particles is merely a *possible* one, considered because of our ignorance of the exact state of a particle, then Statistical Veridicality tells us that the degree of belief which the theory warrants in the proposition that the particle has a given attribute (such as being in region *V*) will equal the degree of belief which the theory warrants that an observation of it will lead to that result.

The measurement of spin is not statistically veridical. Consider an eigenstate for the YNO of whether the spin is up with respect to direction *a*. And let us observe whether or not it is also up with respect to direction *b*. I have taken the distribution of spins to be (approximately) uniform in the 'up' hemisphere with respect to *a*. So the proportion of systems in the collection which are also 'up' with respect to *b* will be $1 - \theta/180°$ where θ is the angle between *a* and *b*. But the predicted proportion of observations which give an up result turns out to be $(\text{cosine}\, \theta/2)^2$ (Messiah, 1962, p. 546). This coincides with $\theta/180°$ only for the special cases $\theta = 0°$, 90° and 180°. (That is, if *b* is parallel to *a*, at right-angles to *a*, or opposite *a*.) In no other case do we have Statistical Veridicality.

There is one further question concerning measurement theory. Does the converse of Weak Veridicality hold? That is, are the only quantum states for which Yes is the true answer to a YNO the eigenstates for that YNO? Since a quantum state is now considered to describe a whole collection of particles, this question is ambiguous. When we say that the Yes answer is true, do we mean it is true of *all* the members of the collection, or of some particular member (say the only actual one)? If we took it the latter way, then the converse of Weak Veridicality would not hold. Consider a state giving a probability 1/2 of a Yes answer to a position YNO. That answer would be true of about half the members of the collection. However, if we take it the former way, then we have something very close to the converse. For a large collection it will be extremely improbable that the Yes result is true of all members unless the state approximates an eigenstate, in the sense that it gives a very high probability of a Yes result. In that sense, then, the converse holds approximately.

VII SYSTEMS OF SEVERAL PARTICLES

Formally a system of n particles is like a single particle whose location is in a 3-n dimensional space. Accordingly, I shall say that such a system has a *location polytuple*, namely the n-tuple of the locations of the n particles. The rate of change of this location polytuple specifies, for the system, an occurrent *momentum polytuple*, which is simply the n-tuple of occurrent momenta. We can now adapt the account I have given for one-particle systems. The quantum state will describe a collection of n-particle systems and specify the frequencies for the location polytuples of the systems (and hence the locations of the n particles). There will also be tied relative propensities for the change of the location polytuples of the system.

The propensity for the change of location polytuple of the n-particle system is not always analysable into the propensities of the individual particles to change their locations.[25] So we require a minor conceptual innovation when we consider systems of several particles. Previously I have taken propensities to be properties of particles. We must now either consider a propensity to be a property of the system or to be a relation between the particles. (In the latter case a propensity relation might be characterized just like a propensity property. I leave the details to the reader.) Which should we prefer? First note that if we have a propensity relation between the particles we also have a propensity property of the system. That is because given any relation R between the particles, we can always consider the property of the system of being made up of particles related by R. Call this the *derived* property of the system, $\hat{R}$. Conversely, given any property P of the system we can always consider the relation between the particles of being disjoint parts of a system with that property. Call this the *derived* relation P^+ between the particles. However $(\hat{R})^+$ is not always the same relation as R. For example, suppose we start off with the anti-symmetrical relation *being earlier than* as R, then $\hat{R}$ is the property of having two parts one of which is earlier than the other. Hence $(\hat{R})^+$ is the relation which two things are in if they are parts of something made of an earlier and a later part. That is a quite different relation. For, among other reasons, it is symmetrical.

Do we have, then, a propensity of the system derived from a propensity relation between particles, or a propensity relation between particles derived from a propensity of the system? Consider a system made up of two qualitatively identical particles initially at locations q_1 and q_2. Suppose there is a certain propensity for a change to a system in which the particles are at $q_1{}^*$ and $q_2{}^*$, where $q_i{}^*$ is very near q_i compared to the distance between q_1 and q_2. It is rather plausible to say that there is a propensity of the particle at q_1 to move to $q_1{}^*$, and for the particle at q_2 to move to $q_2{}^*$, rather than simply say that the propensity is for the system to change to one where the particles are at $q_1{}^*$ and $q_2{}^*$. Hence the propensity of the system is best considered to be derived from a propensity relation between the particles.

This, by no means conclusive, consideration, together with the intuition that we should analyse complex systems, where possible, in terms of relations between parts, makes me favour treating the propensity relation as fundamental and the propensity of the system as derived. But not much depends on this decision. Whichever way we decide, we have reached one of the less attractive features of the Determinate Particle Speculation. For not only do we have a striking abandonment of the Dyadic Programme, we have, as far as I can see, *unanalysable* propensities for the complex system, or else *unanalysable* propensity relations of high polyadicity.

We shall have Statistical Veridicality for the location polytuple of the n-particle system and for its occurrent momentum polytuple. This is simply an extension of those results for the one-particle case.

What can we say of systems of many particles with *spin*? For simplicity consider the case of spin one-half particles, and ignore location and momentum. Then the quantum state will in general be a superposition of decomposable states. We assign a distribution of spins to the decomposable states by assuming independence. But for the non-decomposable states we have little guide to the distribution. This is a gap in the proposed speculation.

VIII BELL'S INEQUALITY AND THE TWIN-SLIT EXPERIMENT

I have discussed both Bell's Inequality and the twin-slit experiment in chapter 3. But there are a few additional remarks I should make in connection with the Determinate Particle Speculation.

Given the usual assumptions, Bell's Inequality shows that the simultaneous observation of whether the first particle is spin-up with respect to direction *a* and the second spin-up with respect to direction *b* cannot be statistically veridical. That is, the joint frequencies for the observation do not even approximate the joint frequencies in the collection. For if they did, the correlations should be the same for the actual spins and the observation of spin. Those for the actual spin must satisfy (the Eberhard version of) Bell's Inequality, while those for the observations do not. This argument from Bell's Inequality is independent of any assumption about the precise distribution for spin, provided it is positive. So it serves as a rejoinder to anyone who would try to preserve Statistical Veridicality for spin by rejecting the distribution for spin which I have posited. But, as I argued in chapter 3 (section III), we should not expect Statistical Veridicality for spin anyway.

What conclusion can be drawn from the twin-slit experiment? To save the Dyadic Programme we would have to posit rather special forces which the second slit exerts on a particle going through the first slit, and which it exerts regardless of the precise physical composition of the material surrounding the slit.

I see no need to posit such forces. Instead, I accept that there are non-accidental statistical correlations at a distance which are not due to forces. These will manifest the propensity relations between the particles of the system which govern their change of location. Indeed, that these propensities are relations between particles and not properties of individual particles explains why we have such non-accidental correlations at a distance. That is precisely how we should expect propensity relations to be manifested.

In connection with the twin-slit experiment, Popper has emphasized that the propensities involved are not properties of the particle going through the slit but of the whole system (1982, pp. 152–3). This I take to be a similar rejection of the Dyadic Programme.

Once we reject the Dyadic Programme there can be no strong objection from the twin-slit experiment to the Determinate Particle Speculation. But there may remain a residual worry. This is that the results of the twin-slit experiment fit a wave interpretation so neatly. Does not that show there is something wave-like going on? I reply that there is indeed something wave-like, namely the relative propensities. For a particle with relative propensities has a propensity for each location, not just for its own location. The set of relative propensities, then, is represented by an assignment to each location of some item. Hence it is *formally* like a disturbance to a field. The strength of the field at a given location corresponds to the probability distribution for change of location. Hence wavy behaviour is as much to be expected for relative propensities as, say, for disturbances to the electromagnetic field.

IX THE PECULIAR LAWS GOVERNING DETERMINATE PARTICLES

The Determinate Particle Speculation requires laws (or 'laws') governing its dynamics which depart further from those of classical mechanics than is often thought. It is tempting to think of the dynamics as simply replacing necessity by probability, but as not otherwise departing from the classical theory in which the state of the system at one time causes it to change to a new state at a later time. And if we did not have to posit tied propensities (in order to handle the Redundancy/Interaction Dilemma) we could think of the dynamics that way. But tied propensities can only be specified by describing, among other things, a privileged class of collections. Since our account of the dynamics must make explicit mention of such a privileged class it is not just a matter of replacing necessities by probabilities in the classical account.

Moreover, I proposed an account of propensities in chapter 4 which went some way towards explaining what propensities were. For absolute propensities this account required the fundamental law that the vast majority of collections conform. This account was extended to cover tied propensities by taking the law to be that the vast majority of privileged collections conform. This makes tied propensities less mysterious than they would otherwise be. Hence the need to posit tied propensities itself supports my account of propensities. This in turn emphasizes the novelty of the laws governing the dynamics.

The laws governing the dynamics of systems are of a novel kind, then. To bring out this novelty I distinguish *positive* from *negative* laws, as in chapter 3 (section VII). Positive laws show us what causes what, or what is likely to cause what. Negative laws put constraints on the world, which any causal relations must conform to, but they do not directly specify what is likely to cause what. Newton's Laws of Motion are positive laws. The Conservation Laws are negative ones. Part of the novelty of the dynamics of the Determinate Particle Speculation is that *all* the laws are negative.

Not only are the laws negative, they are holistic. (The two usually go together.) For they do not entail anything about any given privileged collection, only that the vast majority of them conform.

Neither of these novelties should be taken as a decisive objection to the Determinate Particle Speculation. We have no justification for insisting that the laws of nature be positive. Perhaps we have a metaphysical intuition that every change has a cause. But to convert that into an a priori argument for positive laws we would require an extra premiss, namely that the physical sciences will serve to specify what causes what. And to assume that extra premiss a priori is to play Science Says.

Yet another peculiarity of the dynamics required on the Determinate Particle Speculation concerns the Conservation of Energy, or rather its apparent failure to be conserved in the *tunnel effect*, namely the penetration of a potential energy barrier (see Hooker, 1971, p. 266). Consider, for instance, a collection of electrons in an eigenstate for whether the energy is less than some value E_1. By Weak Veridicality all the electrons have kinetic energy less than E_1. Now suppose there is a *potential barrier* at which the beam is aimed. This is a region where there are forces pushing the electrons back. A dense region of positive charge would be an example. The energy classically required to penetrate this barrier is, I shall suppose, much greater than E_1. So we would expect the electrons to bounce back from the potential barrier. Yet sometimes they penetrate it (Messiah, 1961, pp. 96–8). It is as if they 'borrow' energy for a short time.

We should note that on the Determinate Particle Speculation the electrons really do cross the barrier. The only alternative would be to insist that some of them were already on the other side. But this tunnel effect is used to explain alpha radioactive decay (Messiah, 1961, p. 92), and to construct tunnel diodes. In both cases there is genuine penetration of the barrier.

Is there a failure of energy conservation? There would seem to be a failure of the classical principle for individual particles in the collections – unless we allow that the potential energy takes negative values. For the sum of the kinetic plus potential energies cannot be conserved. On the other hand, there is no failure of the quantum theoretic version of the Conservation of Energy, discussed in chapter 3, for that merely tells us that the probabilities of various results of energy YNOs are constant. And there is never a non-zero probability of a Yes result for the YNO of whether the particle has energy greater than E_1. (There is no YNO of observing the energy while the particle is in the barrier.)

How serious is the failure of the classical principle? That depends how seriously we take potential energy. If we think of it as energy actually stored in some field, then the classical principle tells us that the total actual energy is constant. In that case, we have an immensely elegant 'law of nature', one we should not lightly discard. On the other hand, if we are realists about the forces, rather than the fields which could be used to replace forces, then considering potential energy is merely a useful device for keeping track of changes in kinetic energy, which is then the only actual energy. In that case, the classical Conservation of Energy is not worth fighting for.

The upshot is that if we accept that *all* forces are due to fields, then we have another significant, but not overwhelming, defect in the Determinate Particle Speculation.

X ASSESSING THE DETERMINATE PARTICLE SPECULATION

I shall now list some of the good and bad points of the Determinate Particle Speculation. Let us start with the good. It handles Schroedinger's Cat with ease. In 50 per cent of cases the cat is exploded, in 50 per cent it is not, but there is no limbo between being exploded and being intact. Its second great advantage concerns the Problem of the Two Dynamics. On the Determinate Particle Speculation the only problem is the Threat of Statistical Incoherence. I assume that this threat can be averted, so that, *for a large collection of systems* interacting with a piece of apparatus, the results of the Schroedinger dynamics are indistinguishable from the results of the von Neumann dynamics. The remaining problems concerning the two dynamics were all based on the claim that the Schroedinger dynamics made different predictions for individual systems than the von Neumann dynamics. But on the speculation I am here defending, that claim has a false presupposition. The Schroedinger dynamics does not directly constrain individual systems at all. For all it (directly) constrains are quantum states. And on the speculation I have provided, quantum states are not states of individual systems, but rather of large collections.

Before I consider the genuine disadvantages of the Determinate Particle Speculation, I shall note two spurious ones. One is that it requires correlation at a distance; but that is no defect. Another spurious disadvantage is that we need negative probabilities. But I have argued that negative probabilities are quite acceptable given my account of propensities.

I now consider the genuine disadvantages. The first is that while the conceptual innovations involved in tied propensities and the novel dynamics are *legitimate*, the account given turns out to be more *complicated* than we would have hoped. A law to the effect that the vast majority of privileged collections conform, although easy to state, is more complicated than a constraint directly on individual systems. Moreover, we would rather have avoided unanalysable propensity relations of high polyadicity (or, alternatively, propensities of complex systems).

Another genuine disadvantage is that the speculation is incomplete. I have not been able to say what the probabilities for the spins are for systems of many particles when they are not in decomposable states. Spin remains mysterious.

Finally the violation of the *classical* Principle of Energy Conservation is a genuine disadvantage if and only if we consider that all forces are due to fields about which we are realists and which store real energy. Otherwise this is another spurious disadvantage.

NOTES

1 Suppose the joint probability distribution is given by $F(q,p)$ and the observable is a function $g(q,p)$ of position and momentum. And suppose the corresponding operator is G. Then if Statistical Veridicality holds for that observable, we should have Cohen's condition d_1 (Cohen, 1966, p. 317), namely:

$$\langle G\rangle = \iint g(q,p)F(q,p)\,dp\,dq.$$

Here $\langle G\rangle$ is the mean value of the results of the observation. If Universal Statistical Veridicality held, we should also have this result for a function K of $g(q,p)$. That is, we should have d_2, namely:

$$\langle K(G)\rangle = \iint K(g(q,p))F(q,p)\,dq\,dp$$

Cohen's conditions (b) and (c) express Statistical Veridicality for position and momentum respectively. Cohen then shows that not all of (b), (c), (d_1) and (d_2) can be satisfied.

2 Kochen and Specker assume that in a hidden variables theory the expected value assigned to an observable A is given *both* by the usual quantum formalism and by a formula labelled (2) (1967, p. 67). But the quantum formalism gives the expected results of observation, whereas (2) is the mean value for the observable prior to observation. Equating the two amounts therefore to assuming Statistical Veridicality.

3 Curiously Redhead (1987, p. 123) talks about making a further 'plausible assumption'. This is the assumption that for any hidden state there is some observable which is non-zero for that state. If we take the formalism to be in terms of YNOs this is not an extra asssumption at all. Take any YNO, A, then either A or $\bar{A}$ will be observable, which is non-zero for the state.

4 This *minimal instrumentalist interpretation*, as Redhead calls it (1987, p. 44) is indeed interpretation not speculation. I take it to be part of quantum theory itself, and it is implicit in my exposition of chapter 2.

5 See Redhead (1987, p. 45). Redhead misleads us a little, when he writes 'View A says that QM is not really mysterious at all. It is just glorified statistical mechanics.' What Redhead calls a hidden-variable interpretation is indeed a 'glorified' statistical mechanics. But as he concedes (p. 46), it contains more than just View A. View A is what I mean by the Determinate Particle Speculation.

6 For simplicity, suppose the errors in measuring A are $+\delta$ and $-\delta$, each occurring with the same frequency. Then these cancel out for A. If the actual value of A is a then the observed values of A will be $a+\delta$ and $a-\delta$. Hence the observed values of $f(A)$ will be $f(a-\delta)$ and $f(a-\delta)$. These will 'cancel out' only if $\frac{1}{2}f(a+\delta)+\frac{1}{2}f(a-\delta)$ is a very good approximation to $f(a)$. Otherwise there will be a distortion to the mean value of the observable, contrary to Statistical Veridicality. Now by elementary calculus if $f(x)$ is suitably differentiable, $\frac{1}{2}f(a+\delta)+\frac{1}{2}f(a-\delta)=f(a)+\delta^2f''(a)+$ terms of order δ^4. It is reasonable to require accuracy to the order δ^2, and hence that $f''(a)=0$. Since this is to hold for all a, the function is linear.

7 Degrees of belief are taken to cover both beliefs of varying confidence and disbeliefs of varying confidence. So having a degree of belief that p does not entail believing that p.

8 For my long story see Forrest (1986, chapter 4).

9 It should not be assumed without question that there is a *unique* warranted degree of belief. Uniqueness would have to be part of some hypothesis.

10 Popper says that he has abandoned a frequency interpretation in favour of propensities (1982, p. 104). This suggests that he might accept this variant. But an alternative proposal I shall be developing would be equally in the spirit of Popper's remarks.

11 Here, as elsewhere, the probabilities only approximate the frequencies. The theory tells us what happens in large collections, and is more accurate the larger the collection.

12 If we know, for instance, that the system is in its lowest energy level then this could specify various probabilities for the various YNOs.

13 Moreover positing negative mass accounts for the spin as literally due to rotation (see chapter 2, note 19).

14 The Many Worlds Theory is usually associated with the Everett interpretation of quantum theory (1957). However, Everett's proposal was, initially at least, that the total state be considered a superposition of states corresponding to worlds. I have already discussed this when considering the Problems of the Two Dynamics. It can be interpreted as like the heroic metaphysics I there mentioned. However, de Witt has tended to treat the Many Worlds interpretation as a theory requiring a collection of worlds (sub-universes) (Everett et al., 1973).

15 Compare van Fraassen's *modal interpretation* (van Fraassen, 1972, pp. 335–43).

16 Given the standard formalism, this is a trivial result. We represent the pure and mixed states by trace class operators on a Hilbert space (Beltrametti and Cassinelli, 1981). Then the Schroedinger dynamics is represented as follows. For each quantity of time t there is a unitary operator U_t. If T is the trace class operator representing the state at one time, $U_t\, T\, U_t^{-1}$ will represent it at a time t later. Mixing states represented by T_1 and T_2 in the proportions λ to $(1-\lambda)$ results in the state $\lambda T_1 + (1-\lambda) T_2$. Since U_t is linear, it follows that $U_t(\lambda T_1 + (1-\lambda) T_2) U_t^{-1} = \lambda U_t T_1 U_t^{-1} + (1-\lambda) U_t T_2 U_t^{-1}$. Hence the Schroedinger dynamics preserves mixture.

17 Let the joint probability *density* be $P(q,p)$. Let $F(q) = \int P(q,p)\,dp$ be the distribution for location. Let $G(p,q) = P(q,p)/F(q)$, if $F(q) \neq 0$. Then $G(p,q)$ is the density for the relative propensities. Now Wigner (1932) provides an equation of the form:

$$(1) \quad \partial \mathrm{P}/\partial \mathrm{t} + (1/m)\mathrm{p}.\mathrm{grad}_{\mathrm{q}}\mathrm{P} = \mathrm{V(P)}.$$

Here V is an operator involving derivatives only with respect to p, and $\mathrm{grad}_{\mathrm{q}}$ is a differential operator involving derivatives only with respect to q. From (1) we find that:

$$(2) \quad \partial \mathrm{G}/\partial \mathrm{t} + (\mathrm{G}/\mathrm{F})\partial \mathrm{F}/\partial \mathrm{t} + (1/m)\mathrm{p}.\mathrm{grad}_{\mathrm{q}}\mathrm{G} + (\mathrm{G}/m\mathrm{F})\mathrm{p}.\mathrm{grad}_{\mathrm{q}}\mathrm{F} = \mathrm{V(G)}.$$

Hence $\partial \mathrm{G}/\partial \mathrm{t}$ is independent of F only if $[(1/\mathrm{F})\partial \mathrm{F}/\partial \mathrm{t} + (1/m\mathrm{F})\mathrm{p}.\mathrm{grad}_{\mathrm{q}}\mathrm{F}]$ is. And that will occur only for highly restricted choices of F, namely those of the form $\exp(at + bx + cy + dz)$.

18 This is not the only possible formula, but it is the simplest. So I shall assume it is correct, if any is.

19 The current at q is given by $\underline{J} = \underline{1/m} \int P(q,p)$. By partial integration this becomes: $J = (1/2\mathrm{i}m)(\Phi \mathrm{grad}\bar{\Phi} - \bar{\Phi}\mathrm{grad}\Phi)$, a formula given in Messiah (1961, p. 121). If we write $\Theta(q)$ for $\mathrm{Arg}\Phi(q)$, then $J(q) = (1/\mathrm{m})|\Phi(q)|^2$ grad Θ.

20 Since $J = (1/m)|\Phi(q)|^2$ grad Θ, J and the probability distribution for location jointly specify grad Θ wherever the probability density is

non-zero. The required assumption is that the region where the probability-density is non-zero is connected. For then the probability for location together with the current specify $|\Phi(q)|$, and $\Theta(q)$ up to an additive constant k. Hence $\Phi(q)$ is specified up to a multiplicative constant e^{ik} and so the state is specified.

21 As in note 21 to chapter 2, we can associate to each (pure) quantum state u for a particle of spin one-half, a state u^* for a spinless particle and a spin direction $\sigma(q)$ for each location q at which the probability density for location is non-zero. Presumably we can use u^* in Wigner's formula, to give us a joint distribution $P(V, W)$ for location and momentum. If we take the spin for particles at q to be uniformly distributed around the up hemisphere with respect to $\sigma(q)$ we shall then have a distribution for location, momentum and spin.

22 Suppose, as in note 21 to chapter 2, a pure quantum state is given by a pair of complex functions $\langle\Phi_1(q),\ \Phi_2(q)\rangle$. Then if we ignore the location we can consider the state of a particle of spin one-half to be represented by a pair of complex numbers $\langle z_1,\ z_2\rangle$. In that case, the associated spin direction σ is vector with co-ordinates $\langle z_1\bar{z}_2 + \bar{z}_1 z_2, -i(z_1\bar{z}_2 - \bar{z}_1 z_2), |z_1|^2 - |z_2|^2\rangle$.

23 See chapter 2, note 19.

24 Consider any set W of momenta. Let Sp be the whole of space. Then: $\Pr(Sp, W) = \int \Pr_q(W)\mathrm{P}(q)\,dq$, where $\mathrm{P}(q)$ is the probability density for position. Given Wigner's distribution, $\Pr(Sp, W)$ is the integral over W of $|\Phi(q)|^2$, and so is the probability that the result of observing momentum is in W. But $\int \Pr_q(W)\ \mathrm{P}(q)dq = \int \mathrm{Prop}_q(W)\mathrm{P}(q)dq$. Since $\mathrm{P}(q)$ is the density for a collection in the privileged class, the propensities are manifested by the appropriate occurrent momenta. Hence $\int \mathrm{Prop}_q(W)\mathrm{P}(q)dq$ is (to a good approximation) the proportion of members of the collection with occurrent momenta in W. Hence we have Statistical Veridicality.

25 Suppose it were. Then the joint probability distribution for the momentum of the n-particles would have to be given as a function of the probability distributions for the momenta of the individual particles. Where the state is decomposable we have probabilistic independence for the momenta. However, for suitable superpositions of decomposable states we do not. Indeed for two spinless particles we have wave functions $\Phi(q_1, q_2)$, which are antisymmetrical, that is: $\Phi(q_2, q_1) = -\Phi(q_1, q_2)$. (For example, let $\Phi(q_1, q_2) = \psi_1(q_1)\ \psi_2(q_2) - \psi_2(q_1)\ \psi_1(q_2)$.) Their Fourier transforms are also antisymmetric. Hence $\hat{\Phi}(p, p) = 0$. Given independence this would show that the probability density for momentum was always zero, which would be absurd.

6

Waves

A pure state can be described by means of a suitable (complex- or spinor-valued) function. Let us speculate that this function represents a disturbance to a field, that is, a wave.[1] On this speculation, a 'particle', such as an electron, is in fact a wave not a particle. One difference between a particle and a wave is that a particle is located in a finite region, but a wave can have non-zero values everywhere. The wave speculation which I shall develop has been defended by Schroedinger (1952). Such speculations should be distinguished from various *alternatives* to quantum theory based on waves. Among these I include de Broglie's original 1920s proposal (see de Broglie, 1960), which had an important influence on the development of quantum theory. I also include the theory of Bohm and Bub (1966), and of Destouches (1980). But none of these is within the scope of my inquiry.

I THE SIMPLEST CASE

As usual I begin by considering a single spinless particle. I shall describe the basic Wave Speculation. In section IV I discuss two variants on this basic speculation.

A pure state is specified by a wave function Ψ assigning a complex number to each location.[2] That this is a *complex* number introduces a couple of complications. Let us, therefore, begin by considering the very special situation in which the state is represented by a real-valued function which is never negative. So it assigns either zero or a positive number to each location. In this case I speculate that $\Psi(q)$, the value assigned to q, describes the field at q. More precisely, I speculate that there is a set of properties (the *values* or *intensities* of the field at q) which *could be* instantiated at q, and that they are in one-to-one correspondence with the positive numbers.

If Ψ assigns zero to q then none of these properties are instantiated at q.[3] In all other cases the property corresponding to the number $\Psi(q)$ is instantiated at q. In addition, I speculate that some of the mathematical structure of the positive numbers represents relations between the field-values. Thus I speculate that there is a dyadic[4] operation on these properties which I shall call *summation*. The sum of two field-values is a third field-value. Summation is represented by the addition of the numbers. Again I speculate that for any positive real number r there is a relation which holds between the field-values if the first is represented by a number of magnitude r times the second.[5]

I mentioned that the use of complex numbers introduced a couple of complications. One of these is best discussed in section IV, since it leads to the consideration of a slightly different speculation. I shall here discuss the other, although it may be ignored without much jeopardy to our metaphysical speculations.

Using the Argand diagram (figure 6.1), complex numbers may conveniently be treated as vectors in a two-dimensional plane. If the complex number is z, then it is represented by a vector OP of a certain length $|z|$ called the *absolute value* of z. Unless $|z|$ is zero, this vector makes a certain angle α with the x axis. The angle α is called the *argument* of the complex number.

We are to assume that the field-values at a given location are in one-to-one correspondence with the (non-zero) complex numbers. The complication is that they are specified not just by a positive real number

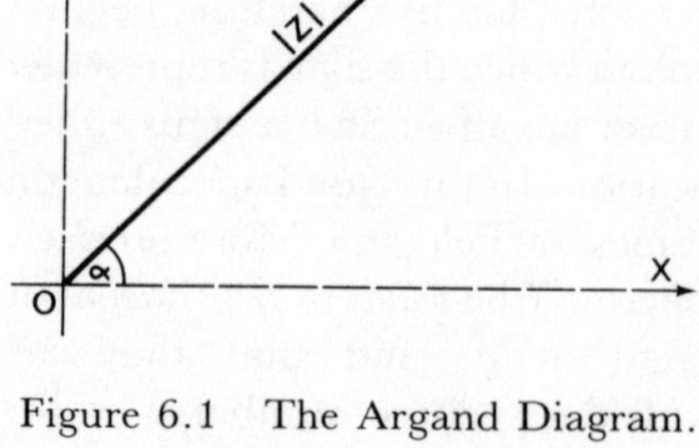

Figure 6.1 The Argand Diagram.

$|\Psi(q)|$, the *amplitude* of the wave, but also by an angle, the argument of $\Psi(q)$, called the *phase angle* at q.

I shall assume, largely for simplicity, that the set of possible field-values for any two points in space is the same.[6] I shall also assume that the same complex number represents the same field-value at the two different points.[7] Both these assumptions could be questioned, but the effect of rejecting them would not, I submit, make the Wave Speculation any less tenable. All it would do is complicate the exposition.

Fortunately the use of the complex numbers does not greatly affect our discussion. I shall assume that the wave function specifies a disturbance to a field. We need not be reminded at each step of the *way* in which it does this. In deference to the first complication I shall write the amplitude of the wave as $|\Psi(q)|$. But if we simplified matters by insisting that Ψ is always a non-negative real number, then the amplitude would just be $\Psi(q)$.

The 'particle' is entirely in the region V if the state is an eigenstate for the corresponding YNO and hence if $\Psi(q) = 0$ outside V. So we interpret a wave's being in V as the disturbance being contained in V. This is the natural way of giving a wave a location. Now, for all I know actual waves are never contained in finite regions and the disturbance is never zero. But in many cases there will be very little disturbance outside V. So we shall have approximations to waves contained in finite regions. And if the region is small the wave is then particle-like. One crucial difference between a wave contained in a small region and a particle is that the former but not the latter has a capacity to spread, taking up indefinitely large volumes.

On the speculation which I am here expounding, a 'particle' is spread out. So we should expect it to have a varying *mass density*, which is zero outside the region occupied by the 'particle'. I speculate that this mass density is proportional to the square of the amplitude of the wave. As a consequence, the probability of a Yes result for the YNO of whether the 'particle' is in region V will be equal to the proportion of the total mass in V. The neatness of this result is one reason for assuming the mass density is proportional to the *square* of the amplitude, not, say, the amplitude itself. Another reason is that in the theory of electro-magnetism, the *energy* density is proportional to the square of the amplitude of the wave, and we can take the mass to be proportional to the energy by Einstein's celebrated equation $E = Mc^2$.

We have motivation, then, for speculating that the mass density $\mu(q)$ is proportional to the square of the amplitude of the wave $|\Psi(q)|^2$. I shall adopt the convention that the constant of proportionality is one. This is contrary to the standard convention that the integral of $|\Psi(q)|^2$ is one.[8] For on the convention I propose this integral will be the mass of the 'particle'.

One advantage of this non-standard convention is that it reminds us that the field intensity is a physical quantity in *units* (of the square-root of mass density), not just a number. We have then: $\mu(q) = |\Psi(q)|^2$. Using this formula we can find the centre of mass of the wave.[9] This is a determinate location *associated* with the wave. It equals the expected value for the results of measurement of position. Hence we have not merely Weak Veridicality for position but also Mean Veridicality.

Let us now consider the von Neumann dynamics. In this case the phrase 'collapse of the wave-packet' is not inappropriate. Suppose the YNO of whether the wave is in region V is made. Then the probability that the wave will come to be in an eigenstate (i.e. that it will come to be contained in V) is equal to the proportion of the total mass in V. I interpret this probability as an absolute propensity for the wave to contract (if necessary) so as to come to be in V.[10] Thus the same property is described occurrently as a proportion of mass and non-occurrently as a propensity for contraction (see chapter 4, section IV).

The measurement of momentum is not so straightforward, because momentum is primarily a characteristic of particles. We may, however, consider an *associated* determinate (occurrent) momentum, namely the total mass times the rate of change of the centre of mass of the wave. According to Ehrenfest's theorem this equals the expected or mean value of the results of measurements of momentum (see Messiah, 1961, p. 218). We have, then, Mean Veridicality for momentum. And in an eigenstate for the YNO of having momentum in set W, the associated determinate momentum will indeed be in W. So if what we are trying to measure is the associated determinate momentum, we have Weak Veridicality as well.

The associated determinate momentum is one analogue of the occurrent momentum of a particle. A second analogue is the rate of transfer of mass across a small plane at a given point q. This will be specified by the *current* $J(q)$ at q,[11] which is zero if the mass density $\mu(q)$ is zero. It will be given by a formula similar to that for the propensity current considered in the previous chapter.[12]

The total rate of mass transfer, given by the integral of $J(q)$ over space, turns out to be equal to the associated determinate momentum. This leaves open the possibility of taking the current J as more fundamental than the associated determinate momentum, even though the latter is being directly measured. I shall discuss this possibility in section IV.

What of the converse of Weak Veridicality? It is easy to check that it holds for location. But if we interpreted the quantity we are trying to observe in momentum observations as the associated determinate momentum, then we do not have the converse of Weak Veridicality for momentum. For, among other reasons, the associated momentum will be determinate and no wave is an eigenstate for a determinate value of momentum.

II POLYWAVES

A single particle, in the Particle Speculation, has been replaced in the Wave Speculation by a wave. What happens to a system of several particles? For simplicity, let us concentrate on two spinless 'particles'. The wave function is now a function of two locations $\Psi(q_1, q_2)$. If the two 'particles' are of different kinds then sometimes the state is decomposable and we have $\Psi(q_1, q_2) = \Psi_1(q_1) \times \Psi_2(q_2)$. If that were always the case, we could take the system to consist of two waves, one represented by Ψ_1, the other by Ψ_2. But in general the wave function $\Psi(q_1, q_2)$ is not decomposable into two wave functions $\Psi_1(q_1)$ and $\Psi_2(q_2)$.

Far from being a reason to reject the Wave Speculation, this problem is a reason for greater boldness in our speculation. I speculate that a 'two particle' system is a system of *dyadic* waves. Quite generally, I posit *polywaves*, which are disturbances to polyadic fields. The familiar monowaves (monadic waves) are assignments to each location of some member of the set of possible field-values for that location. Likewise an n-adic polywave is an assignment to each ordered n-tuple of locations of a member of the set of possible field-values for that n-tuple of locations. The integer n is just the 'number of particles' in the system. And the possible field-values are n-adic relations.

A polywave has a location in a 'region' U, consisting of a set of n-tuples of ordinary locations $\langle q_1, \ldots, q_n \rangle$, just in case the field disturbance is zero for n-tuples outside U. In particular, if U consists of all the n-tuples of locations $\langle q_1, \ldots, q_n \rangle$ such that q_k is in V_k, then a polywave is in U just in case the field disturbance is zero unless q_k is in V_k, for $k = 1, \ldots, n$. With this natural definition we ensure that the measurement of location is weakly veridical.

When I considered monowaves I associated a mass density function $\mu(q)$ to a monowave, using the formula $\mu(q) = |\Psi(q)|^2$. For polywaves we have to think of the mass as dependent on the disturbances to the field. So we require an assignment of a density $\mu(q_1, \ldots, q_n)$ to each n-tuple of locations.

By analogy with the one-particle case I posit a density $\mu(q_1, \ldots, q_n) = |\Psi(q_1, \ldots, q_n)|^2$. Let M be the integral of μ with respect to all n locations. Suppose Ψ is decomposable into $\Psi_1, \ldots, \Psi_n$ with mass densities $\mu_1, \ldots, \mu_n$ respectively. Then $\mu(q_1, q_2, \ldots, q_n) = \mu_1(q_1) \times \mu(q_2) \times \ldots \times \mu_n(q_n)$. So $M = m_1 \times m_2 \times \ldots \times m_n$, where m_i is the integral of μ_i. Hence M has units of mass to the power n, and μ has units of mass per unit volume all to the power n. So it is not a straightforward mass density.

The quantity M depends only on the wave function. For a system of n 'particles' all of the same kind, we would assume that the m_i are all equal. In that case each of the m_i is the nth root of M. But in general the m_i are not determined by the wave function alone. For we could multiply each of them by a positive constant a_i, provided $a_1 \times a_2 \times \ldots \times a_n = 1$. Hence the m_i are determined by M (and hence Ψ), *together* with the ratios of the masses of the different kinds of 'particle'. These ratios affect Schroedinger's equation.[13] So there are properties of the polywaves described by these ratios, in addition to the properties described by the wave function.

Each of the $\mu_k(q_k)$ define a centre of mass – that for the kth 'particle'. The measurements of location for the kth 'particle' will, on repetition of the measurements, approximate the kth centre of mass. And the associated momentum for the kth 'particle' is obtained by taking the rate of change of the kth centre of mass and multiplying by m_k. The mean for the measurement of the momentum of the kth 'particle' will, on repetition, approximate the associated determinate momentum for the kth particle. Thus we have Mean Veridicality for both position and momentum, provided these are characterized as the associated determinate position and momentum.

Polywaves might seem perplexing. The need to posit polywaves may well have been an important factor in the rejection of speculations like mine in the 1920s.[14] I concede that polywaves are *initially* perplexing. However, their rejection was premature, and we can learn to love them.

Part of the apparent difficulty of a polywave speculation is the common distaste for relations. To some extent this is just an unwarranted bias in favour of properties. But in this case, there is a further worry. The relations have to be between separated points; it thus seems they have no location. But these relations are constitutive of the stuff out of which the whole physical world is made. How can that be? I am not convinced that lack of location for relations would be a genuine problem. But I need not discuss this, for I reject the claim that the relations have no location.

Different kinds of entity are located, if they *are* located, in different ways. To take a simple example, an extended object has a single but extended location, while a point-particle has a point location. Again properties are located in a way appropriate to properties. For if there are many instances of a given property, say, *being an electron*, then, that property has many locations. Notice that if some or all of the instances have extended locations (as for the property of being a hydrogen atom), then the property itself has many locations each of which is extended. These could even overlap. The possibility of overlap shows that having many locations is not the same as having one location with many parts.

I claim that relations have locations but in a way appropriate for relations. A relation, like a property, can have many instances. But each instance will relate a number of objects (perhaps in some specified order). So for each instance we have a number of locations, perhaps in a specified order and perhaps themselves extended. *That* is the way a relation is located, if it *is* located. Formally, we could define the location of a n-adic relation R as a set of n-tuples of regions of space. Each n-tuple corresponds to an instance of the relation, and each region in the n-tuple is the location of an object related in that instance of the relation.

Granted that the above is the way relations are located, if they *are* located, then we can ask of a given relation whether it is located or not. The ones required on the polywave speculation are indeed located. Therefore this source of perplexity has been removed.

I suspect, however, that much of the perplexity derives from the *holistic* character of a polywave theory. For the whole physical universe will just be one big polywave. If I am right about this source of perplexity, there is comfort in the way that actual polywaves can be analysed in terms of (regrettably non-actual) monowaves. Recall that a polywave of polyadicity n is specified by (a) its associated wave function and (b) the associated kinds and numbers of the 'particles' and in particular the associated mass of these 'particles'.

The kinds and numbers of 'particles' for the polywave will be specified by an integer for each kind, (e.g. 10 electrons, 11 protons, 8 neutrons). Let us call that specification simply the *kind* of polywave. Then I propose that there are a few basic operations on possible monowaves and polywaves.

(i) *Boson composition:* From n monowaves (represented by $\Psi_1, \ldots, \Psi_n$) all of the same kind, if that kind is a boson, we can form their composite, which will be specified by the kind (namely n of that kind of boson) and its wave function, which is the symmetrical product of the $\Psi_1, \ldots, \Psi_n$.[15]

(ii) *Fermion composition:* From n monowaves of the same kind, if that kind is a fermion kind, we form a composite as in case (i) except that we form the antisymmetrical product of the wave functions.[16]

(iii) *General composition:* From polywaves (or monowaves) each of which corresponds to a number of the same boson kind, or a number of the same fermion kind, we can form a composite, provided the fermion and boson kinds out of which they are composed differ. The resulting wave function is simply the product of those for the components.

(iv) *Superposition:* Given any two polywaves of the same kind, represented by wave functions Ψ_1 and Ψ_2, and any ratio of complex numbers there is a polywave of that kind (a superposition of the two polywaves) whose wave function is of the form $\alpha\Psi_1 + \beta\Psi_2$, where α and β are in the specified ratio.

Provided we are prepared to be realists about these operations we can treat any polywave as a superposition of composites of possible monowaves. However this decomposition will not be unique.[17] Nor can we think of the decomposition as being into *actual* monowaves. Partly this is because of the non-uniqueness, and partly because superposition is a way of specifying a wave which is in some way *between* two given waves, rather than being *composed* of those waves.

This account of the structure of polywaves should remove the worry due to their holistic character. Admittedly, it suggests further problems, which I shall discuss in section V, but we have shown that polywaves are not as perplexing as they might initially appear.

The need for polywaves may be compared to the need for propensity-relations in the previous chapter. In both cases we have a departure from a tradition of thinking in terms of the *properties* of particles. Instead, in the vanguard of Relation Liberation, we posit analogous *relations*.

III WAVES WITH SPIN

For a single 'particle' with spin one-half, the wave function is spinor-valued. That is, the set of possible values of the field for a given location is no longer represented by complex numbers but *spinors*. The field is therefore complicated by the addition of a 'spin' vector $(1/2)\sigma(q)$ of magnitude one-half assigned to each location for which $\mu(q) \neq 0$.[18]

The average spin for the wave is obtained by multiplying $(1/2)\sigma(q)$ by $\mu(q)/M$ and then integrating over location. I shall assume that this is what we are attempting to measure when we measure spin. It is hard to see what else we could be measuring. I interpret the YNO of whether the spin is up with respect to direction a as whether the average spin is of magnitude one-half and in direction a. (Contrast this with the accounts given in chapters 5 and 7.) This ensures both Weak Veridicality and its converse. For the waves with average spin of magnitude one-half in direction a are precisely those for which $\sigma(q)$ is itself always in direction a. And these can be shown to be precisely the eigenstates for whether the spin is up with respect to a.

For a system of many 'particles' with spin, the polywave will have values in a field represented by a tensor product of spinor spaces. However, the details need not concern us, for we can analyse any such polywave as a superposition of polywaves decomposable into monowaves with spin. And those we have already considered.

IV FIELDS, FLUIDS OR HYPERSPACE?

For a single spinless 'particle' the mass density $\mu(q)$ and the mass transfer current $J(q)$ jointly determine the state.[19] This suggests that we could think of a 'particle' not as a wave so much as a fluid.

Its state at a given time will be given by its density, the current and (if the 'particle' has spin one-half) spin direction.

Now, if we treat the 'particle' as a wave, then it follows that it is *also* like a fluid. For we can define the mass-density and current in terms of the field. Hence, realism about the field implies realism about the mass density and current. However, the question now arises as to whether we should not abandon realism about the field (and so treat the mass density and current as fundamental).

The prima facie case for this Fluid Variant consists of two arguments which I shall now provide. The most obvious I call the Conservative Argument against waves. It is that a continuous substance with variable density, and with currents flowing in it, is a more conservative account of what there is than a field speculation. It might be thought that this is an advantage.

The second argument, which I call the Phase Angle Argument, is based on the second complication arising from the way the wave function is complex-valued rather than having positive real values. It is that if we add a constant to the phase angle at all locations, we obtain a new complex-valued function which represents the same state.

If, as I have previously proposed, we are realists about properties corresponding to both the absolute value of $\Psi(q)$ and the phase angle (i.e. the argument of the complex number $\Psi(q)$), then we shall be committed to the thesis that there are many observationally indistinguishable waves all corresponding to the same quantum state. We should avoid, if we can, positing a proliferation of indistinguishable states. And we can do so if we are realists about the mass density and the current, instead of properties in one-to-one correspondence with the arguments of the wave function.

In spite of this prima facie case, I shall argue against the Fluid Variant, using the Argument from Superposition. This is the argument that superposition is of such fundamental importance in quantum theory that a speculation on which superposition is a simple operation has an advantage over one on which it is not. Now on the basic Wave Speculation we would expect superposition to occur. It is simply the result of taking suitable linear combinations of the field strengths at each point. This is reflected in the representation by means of wave functions, where we multiply by constants and add in order to obtain superpositions. In short, on a *Wave* Speculation it would be surprising if superposition did not occur. However, if we are realists about the mass density and the current instead of the field, then there is no simple account of superposition.

Moreover, I am able to reply both to the Conservative Argument against waves and to the Phase Angle Argument. The conservative appeal of the Fluid Variant was that we already understood what a mass density is and – optimism breaking in – we already understood the mass transfer current. So we did not need to posit unfamiliar entities. However, this appeal is lost in the case of systems of several 'particles'. For in that case, we would be realists about suitable analogues of the mass density and current. And these analogues will have to determine the state, just as in the 'one-particle' case. So we would have to rely on the density $\mu(q_1, \ldots, q_n)$, and n currents, each of which depends on all n locations.[20] This is as much beyond our previous experience as the polywaves themselves; so there is no advantage in treating the mass density and current as fundamental. Indeed the only way of understanding how a mass density could be assigned to an n-tuple of locations is by taking it to be the mass associated with the energy of a relational field, which brings us back to polywaves.

In reply to the Phase Angle Argument, I say that we should not be embarrassed about positing observationally indistinguishable states, *provided* the account obtained is more satisfactory in other respects. And the Argument from Superposition shows that we *do* obtain a simpler account.

I have argued that the treatment of a 'particle' as a fluid is not to be considered fundamental. That a 'particle' can be so treated depends on its wave-like character. I now turn to another variant of the Wave Speculation. This, the Hyperspace Variant, is based on the familiar tactic of embedding ordinary space in some suitable hyperspace. Since I shall be rejecting this variant, it suffices to consider the simplest possible case. So let us ignore spin. Furthermore, let us pretend that space has two fewer dimensions. On the basic Wave Speculation, then, we assign values to a field at each point on a line. On the variant now being considered we think of space itself not as a line, but as a cylinder.

Each point P on the cylinder is specified by a point Q on the line and an angle α (see figure 6.2). If the 'particle' is given by the function $\Psi(q)$, where q is the co-ordinate for the line, there is a curve on the cylinder given by assigning the angle $\alpha = \text{Arg}\Psi(q)$ to the point q along the line. For example, if $\text{Arg}\Psi(q)$ is proportional to q, this curve will

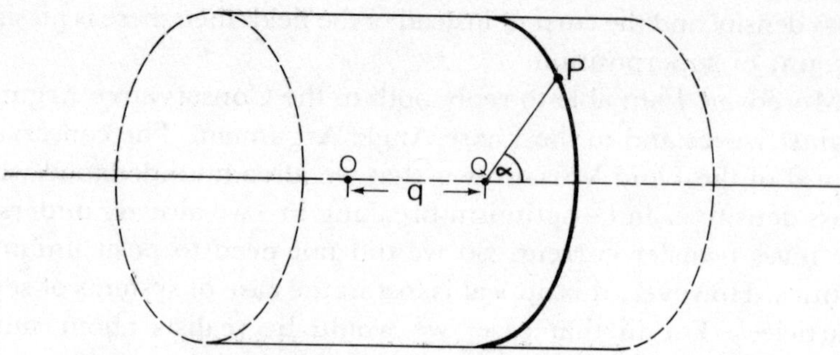

Figure 6.2 Cylindrical space.

be a helix. Then the 'particle' corresponds to a field disturbance which is zero away from this curve and has intensity $|\Psi(q)|$ at the point on the curve given by the point Q on the line.

The prima facie advantage of this variant is that we can explain why adding a constant to the phase angle results in an observationally indistinguishable state. The explanation is that the diameter of the cylinder is very small compared to the observer. So space looks to the observer to be a line not a cylinder. As a consequence, rotating the wave by an angle around the cylinder will not be noticed by the observer, any more than would the displacement of the wave a very small distance along the cylinder.

I have outlined the Hyperspace Variant, because it may well be one that has occurred to the reader. For a 'particle' with spin one-half and with the two stolen dimensions returned, we would have a six-dimensional space with three dimensions being curved around in a space too small to be noticed by us gross observers.

I reject the Hyperspace Variant. First, its prima facie advantage vanishes when we consider systems of more than one particle. On the basic Wave Speculation a 'two-particle' system is a dyadic field whose values are in one to one correspondence with the complex numbers. As in the 'one-particle' case, the quantum state underdetermines the field because adding a constant phase angle gives us the same quantum state. That is the full extent of the underdetermination.

If we think of a 'particle' as a wave on a cylinder-space, then a 'two-particle' system will, presumably, be a dyadic wave on cylinder-space. So an intensity is assigned to the quadruple consisting of two points on a line and *two* angles. But the quantum state assigns to each pair of points an intensity and a *single* angle. There is no obvious way in which the waves correspond to the quantum states. We could, for instance, insist that the field is zero unless the average of the two angles equals the single angle given by the quantum state. But, however we do it, the quantum state will underdetermine the wave by an amount which increases as the number of 'particles' in the system increases. Hence, as regard to indistinguishable hidden states, this variant is worse off than the basic wave speculation. In addition, the Argument from Superposition, which I need not repeat, provides a positive reason for preferring the basic variant. We should, therefore, stick with it.

V POLYWAVE ONTOLOGY

We tend to prefer particle theories, other things being equal. But is this preference based on right reason, or is it mere nostalgia for little gritty granules? More precisely, I ask whether the polywave speculation is intrinsically less plausible than a particle speculation. By *intrinsic* implausibility I mean that which we judge in abstraction from quantum theory itself. It is quite possible that a speculation which is intrinsically less plausible could none the less be accepted because this disadvantage is compensated for by its ability to explain, or otherwise fit in with, quantum theory. None the less if polywaves are much less intrinsically plausible than particles, then the prospects of sufficient compensation are not good.

I shall concede that realism about polywaves has some intrinsic disadvantages when compared with realism about particles, but these are fairly minor. Most of the apparent appeal of particles is familiarity and nothing else. In order to compare realism about particles with realism about polywaves, I need a *generic* account of which these two realisms are *species*. The account I shall give involves properties and relations. The basic proposal is that there are various properties and relations, and that these belong to various items (which may or may not themselves be properties and relations). As I indicated in chapter 1, I deny that such claims about properties and relations are especially controversial.

My strategy will be to begin with realism about particles. There are several different accounts one can give of particles. I shall appeal to the reader to grant me that one of these accounts is no less plausible than the others, or at least, not much less plausible. I shall then compare that selected account with realism about polywaves. Polywaves do not come out of the comparison too badly. They are not much less intrinsically plausible than the selected account of particles, which, in turn, is as plausible, or nearly as plausible, as any other version of realism about particles. I conclude that polywaves are not, to borrow Quine's phrase, *entia non grata*.

I begin by stripping the concept of a particle of one facet which is not required in particle speculations, and which might even make particles less plausible than polywaves. It is their *diachronic unity*. If particles have diachronic unity then we may speak of the *same* particle at one time as at another. And this sameness is not just a matter of spatio-temporal continuity. For consider 'two' cases of three particles[21] of the same kind which hit each other.[22] In the first case they bounce back the way they came (figure 6.3). In the second case they bounce off at an angle of 120° (figure 6.4).

Our ordinary concept of a particle is such that we do distinguish these two possibilities. Yet in both cases we have spatio-temporal continuity. It seems we need some account of what is the difference between the two situations. Perhaps a causal account can be given; but at the least there is a problem. However, I think we can operate perfectly well with a weakened concept of a particle in which we ignore

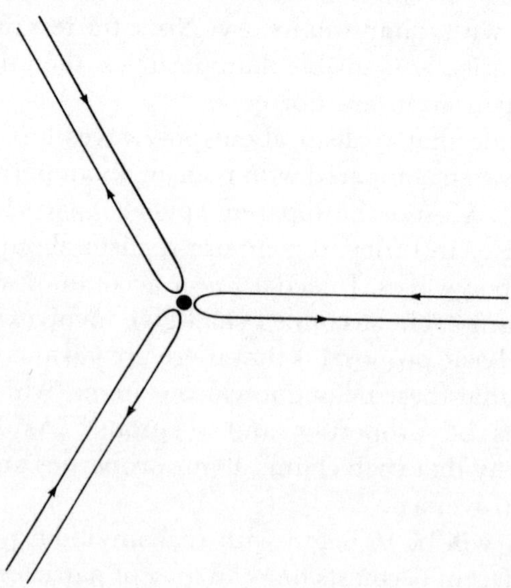

Figure 6.3 Three particles bouncing back the way they came.

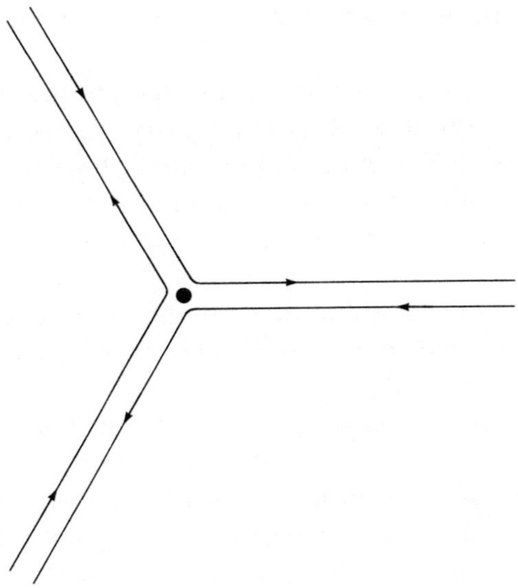

Figure 6.4 Three particles bouncing back at 120°.

diachronic unity. In that case, we say that, for all we know, there is no difference between the 'two' cases: there is just the one case.

A second unity which we think of particles as having is *synchronic unity*. That is, the *one* particle has various different properties and stands in various relations. Conceptually we can distinguish between two particles which happen to have the same location, one with mass but no charge, the other with charge but no mass, on the one hand, from a single particle with both mass and charge, on the other. Some account is needed of this distinction, if we endorse it. Once again if this is in any way embarrassing we could strip synchronic unity away from our concept of a particle. I submit that particle speculations require only such a stripped-down concept of a particle, which although less familiar is, precisely because less burdened with unities, at least as plausible as the common-sense concept of a particle. At this point I speculate that a stripped-down particle is nothing more than the occurrence of various properties at various locations and of various relations between locations. There are alternative speculations, which do not require realism about locations but attempt to paraphrase talk of locations in terms of spatial relations between particles. I think such attempts are doomed to failure. However, even if they succeed, the alternatives are not so very superior to the scheme described above.

I propose, then, a generic account of the physical world, which I call the Location/Property/Relation Theory, according to which the physical world is nothing more than various properties and relations, which belong to, or hold between locations (and, perhaps a hierarchy of further properties of and relations between properties and relations). This is essentially Armstrong's ontology (1978, vol. I, ch. 12). And I appeal to readers to grant that on the particle speculation the Location/Property/Relation Theory is either more plausible or about as plausible as any rival.

I now assume the Location/Property/Relation Theory and compare the plausibility of the Wave and Particle Speculations. Monowaves fit nicely into the Location/Property/Relation Theory. The actual field strength at each point is located precisely at that point. Indeed a particle is very like a monowave, unless we bring back synchronic and diachronic unity. The only difference seems to be that, on a particle speculation, the set of locations where some property belongs consists (at a given time) of discrete small regions (or even points) separated in space. Whereas for a monowave this region will typically be a large volume, and could be the whole of space.

An n-adic polywave consists of various n-adic relations corresponding to actual field strengths and holding between the locations $q_1, \ldots, q_n$ (in that order) to which the field is assigned. Surely passing from properties to relations does not result in any decrease of plausibility.

If the comparison ended here, then I would judge the polywave speculation to be intrinsically as plausible as a stripped-down particle speculation and hence, if anything, more plausible than a common-sense particle speculation. Unfortunately there is a drawback with the polywave speculation. It would be unsatisfactory to propose a single unanalysed polywave with a polyadicity equal to the number of 'particles' in the universe. So it is important to analyse polywaves into monowaves. Corresponding to this analysis I require realism about some operations on the possible properties or relations which make up the components in this analysis. There will be the operations of boson composition, fermion composition, general composition and superposition. First, I concede that there is a cost, that is a measure of implausibility, in requiring realism about such operations. Second, as I have already pointed out, the monowaves resulting from the analysis would not be actual ones. Hence we require realism about uninstantiated properties. Thus, the actual polywave involves a relation instantiated by $q_1, \ldots, q_n$, and this relation is in turn obtained by various operations on properties. But these properties will not be instantiated, that is they need not hold of anything. This is often thought to be objectionable.

I conclude that the polywave speculation has only two intrinsic defects when compared with the particle speculation. First it requires various operations on properties and relations which we would not have anticipated. Second, it requires realism about uninstantiated properties, which many find objectionable.

VI WHERE THE WAVE SPECULATION EXCELS

The Wave Speculation, if correct, enables us to understand superposition and hence the difference between classical and quantum theory. Herein lies an advantage over rival speculations. As we have seen, superposition is a very simple operation given the wave speculation. It only requires that a suitable operation on the field strengths be performed at each location. Indeed given the Wave Speculation we would *expect* superposition. So any superselection rules would seem ad hoc. Furthermore, superposition is one of the key differences between classical and quantum theory. (Others are spin and the classification of particles into fermions and bosons.) So on the Wave Speculation we can explain one of the differences between classical and quantum theory. I shall assume that we take particles to be conceptually stripped down. Then this difference is simply that in some circumstances – those in which classical mechanics is approximately correct – the field is negligible outside a small region, so the wave can be treated as a particle.

By contrast, on the Determinate Particle Speculation, superposition is something we just have to accept. That is not to say that superposition presents a special difficulty for that speculation. Given human limitations, our understanding must fail at some point. Sooner or later we come to that which we can accept but not comprehend. The advantage of the Wave Speculation is that, in one area, we can come to this point later rather than sooner.

A second advantage of the Wave Speculation, scarcely independent of the first, is its naturalness. For on the Wave Speculation we take one formulation of quantum theory, the wave function approach, as mirroring reality in the sense that there is a simple correspondence between what the mathematics seems to describe and what we take to be the case. But on a particle speculation getting from the mathematics to the metaphysics is harder work.

Another, but I think lesser, advantage of the Wave Speculation concerns action-at-a-distance. I have already claimed that this is nothing to get in a fuss about. None the less, action-at-a-distance is not something we would have predicted if we only knew that the physical world was made of particles. On the other hand waves are themselves spread out, and polywaves involve relations between distant points. Hence on the Wave Speculation the action is not distant from the wave.

Again, the Wave Speculation has a minor advantage when it comes to the Conservation of Energy. Recall that there were difficuties with the conjecture that energy is conserved for individual particles on the Determinate Particle Speculation. These difficulties do not arise on the Wave Speculation. For there is a natural conjecture concerning the total energy of a wave.[23] And on it, the total energy is conserved in the Schroedinger dynamics. (Since von Neumann dynamics involves an interaction with something outside the system, we should not expect energy conservation under the von Neumann dynamics.)

VII SCHROEDINGER'S CAT AND THE TWO DYNAMICS

The Determinate Particle Speculation excelled when it came to Schroedinger's Cat. How well does the Wave Speculation cope? About as well, I say, as the Determinate Particle Speculation copes with superposition. We are stuck with a wave which is neither that for an exploded cat nor that for an intact cat, but is a superposition of the two. And even if we posit some extra factor tending to put states into eigenstates for location, we would still have this peculiar situation for a little while. Now we can brazen this out. For we do not have the comically conspiratorial situation where an already exploded cat reassembles itself on observation. Rather we have a speculation in which the world has a blurred, spread-out character to it, which we can never observe because observation leads to greater precision. The world *could* be like that.

However, we do run into difficulties with the von Neumann dynamics. It is not that this dynamics itself becomes complicated.

No, the propensities to go into eigenstates for YNOs are governed by simple enough rules. The problem lies in our having two dynamics. In chapter 3 I suggested various ways of avoiding this difficulty. One is to insist that there is some extra factor always present in the dynamics which tends to become significant in cases of observation. Another was to ignore the von Neumann dynamics altogether and insist that, contrary to common sense, our mental states are as much superpositions as the states of our brain. The problem is not insoluble, but the solutions are literally ad hoc, that is they handle this one difficulty and are specifically designed to meet that difficulty. Now being literally ad hoc is not, I think, as bad as the pejorative use of the phrase suggests. But it is a desideratum not to rely on ad hoc solutions to problems.

VIII ASSESSING THE WAVE SPECULATION

I see no insuperable obstacles to the Wave Speculation, as I have developed it. Its advantages are that it enables us to understand superposition, it does not involve action-at-a-distance and it is a straightforward speculation read off from the wave formalism. Its disadvantages include a commitment to novel operations on properties and relations, and to uninstantiated properties. In addition we have to brazen out the problem of Schroedinger's Cat and resort to a literally ad hoc solution to the Problem of the Two Dynamics. Finally, on the Wave Speculation we do not treat a quantum state as just a probability distribution. Hence we cannot give as satisfactory an account of mixture as was available on the Determinate Particle Speculation. We have to treat mixture as a matter of ignorance about the wave (or of having a large collection of systems each of which is a wave). Hence it is somewhat mysterious why two different mixtures of waves are observationally indistinguishable.

NOTES

1 Waves, recall, need not be sinusoidal, that is, wavy. For each location there is a possible set of values for the field. In an undisturbed field all the actual values are zero. A wave is a smooth assignment of values which is not everywhere zero.

2 In chapter 2 I use 'Φ'. I now use 'Ψ'. This is because I had $\int|\Phi(q)|^2 dq = 1$. But $\int|\Psi(q)|^2 dq$ is the mass of the particle. See note 8.

3 Here I speculate that the zero value corresponds to there being no field property instantiated at the point in question. The alternative is to take zero to correspond to a special field property. There is little to choose between these speculations. If we have localized waves, that is ones which are zero outside some finite region, then it is, perhaps, more natural to speculate as I have.

4 For every n-adic operation there is a corresponding (n + 1)-adic relation. Whether we choose to distinguish the operation from the relation is something we need not decide here.

5 This might suggest something I do not want to suggest, namely that the relation is not identical to the number. See Forrest and Armstrong (1987).

6 What is it to be the *same* field-value at one point as another? In flat space there is no problem. But on a curved space we must rely on *parallel displacement* (Boothby, 1975, pp. 319–30) along a curve connecting the two points in order to make the identification. And there the problem becomes serious. For different curves joining the two points can result in different identifications.

7 It might be thought that we could *stipulate* that the same field-values are assigned to the complex numbers. But if Schroedinger's equation is to hold, the assignment of values to numbers is not open for stipulation. Thus suppose we assign the complex numbers in a non-standard way so that those at q are multiplied by a complex function $F(q)$. Then the wave function $\Psi_t(q)$ will become $F(q) \times \Psi_t(q)$. If both $\Psi_t(q)$ and $F(q) \times \Psi_t(q)$ satisfied Schroedinger's equation, then we would have:

$$(1/F)\left(\frac{\partial^2 F}{\partial x^2} + \frac{\partial^2 F}{\partial y^2}\right) +$$

$$(1/F\Psi)\left(\frac{\partial F}{\partial x}\frac{\partial \Psi}{\partial x} + \frac{\partial F}{\partial y}\frac{\partial \Psi}{\partial y} + \frac{\partial F}{\partial z}\frac{\partial \Psi}{\partial z}\right) = 0$$

This holds for every solution Ψ of Schroedinger's equation only if $F(q)$ is a constant (i.e. independent of q).

8 The standard convention would be motivated by the fact that the *propensities* for Yes results for position YNOs are given by the probability density $|\Phi(q)|^2$. But, we are now thinking of $\Psi(q)$ as describing the field at q as an *occurrent* property.

9 It is the location $1/m\int q\mu(q)dq$ where $m = \int|\Psi(q)|^2 dq$.

10 Let the proportion of the mass/energy in V be λ. Then if $\lambda \neq 0$ the new wave will be represented by $\Psi(q)/\sqrt{\lambda}$ in V, and zero outside V.

11 The rate of transfer across a small plane of area Δ making an angle Θ with the vector $J(q)$ will be $|J(q)|\Delta\sin\Theta$ where $|J(q)|$ is the magnitude of $J(q)$.

12 The current $J(q)$ is $\mu(q)$ times the gradient of ArgΨ. ArgΨ is arbitrary up to an additive constant, but this does not affect J.

13 Namely $i\partial\Psi/\partial t = -(1/2m_1)(\partial^2\Psi/\partial x^2_1 + \partial^2\Psi/\partial y^2_1 + \partial^2\Psi/\partial z^2_1) - \ldots - (1/2m_n)(\partial^2\Psi/\partial x^2_n + \partial^2\Psi/\partial y^2_n + \partial^2\Psi/\partial z^2_n) + V(x_1, y_2, z_1, \ldots, x_n, y_n, z_n)\Psi$.

14 According to a referee for this work. I have been unable to check on this.

15 This is $K\,\Sigma_\pi\,\Psi_{\pi(1)}(q_1) \times \ldots \times \Psi_{\pi(n)}(q_n)$ where K is a constant such that the integral of the square of the absolute value equals $m_1 \times \ldots \times m_n$, and where the sum is over all members of the group of permutations of n items.

16 This is $L\Sigma_\pi(\text{sign }\pi)\Psi_{\pi(1)}(q_1) \times \ldots \times \Psi_{\pi(n)}(q_n)$, where L is chosen so the integral of the square of the absolute value equals $m_1 \times \ldots \times m_n$, where the sum is over all members of the group of permutations, and where signπ is the sign of the permutation.

17 Suppose $\Psi(q_1, q_1) = \Psi_1(q_1) \times \Psi_2(q_2)$. Let $f(q)$ be any real valued function of q. Then $\Psi(q_1, q_1) = \Phi_1(q_1) \times \Phi_2(q_2)$, where $\Phi_1(q_1) = e^{if}(q_1)\,\Psi_1(q_1)$ and $\Phi_2(q_2) = e^{-if}(q_1)\,\Psi_2(q_1)$. Hence the same polywave is obtained from quite different pairs of monowaves.

18 As in note 21, chapter 2.

19 At least if we assume that the region in which $\mu(q) \neq 0$ is *connected*, in the sense that there is a smooth curve joining any two points in the region. For $\mu(q)$ determines $|\Psi(q)|$, and $J(q)$ and $\mu(q)$ jointly determine grad(Arg$\Psi(q)$), wherever $\mu(q) \neq 0$. Hence if that region is connected, Arg$\Psi(q)$ is determined up to a constant, which does not affect the state.

20 For a wave function $\Psi(q_1, \ldots, q_n)$ of n locations, we have n gradients of ArgΨ obtained by differentiating with respect to each of the n location variables. And each of these depends on all n location variables. We could obtain currents which depend only on one location variable each, by integrating over the other variables. But then these currents would no longer determine ArgΨ up to a constant.

21 I have three particles not two to avoid making them pass through each other.

22 I understand that Kripke has made a similar point using the example of a spinning sphere.

23 The energy being considered is the non-relativistic energy. In particular, the mass of the wave, which could be thought of as the energy due to the field disturbance itself is not our present concern. (It is constant for a given kind of particle and so does not affect the conservation of energy.) For simplicity, assume that there are discrete energy levels e_1, e_2 etc. And assume that there are orthogonal wave functions Ψ_1 Ψ_2 etc. such that Ψ_1 is an eigenstate of the YNO of whether the energy is e_i. These

Ψ_i must be normalized so that $\int |\Psi_i|^2 \, dq = m$, the mass of the particle. By Weak Veridicality, the energy of Ψ_i is e_1. Now any wave function Ψ can be expressed as a superposition of the Ψ_i thus: $\Psi = \alpha_1 \Psi_1 + \alpha_2 \Psi_2 + \ . \ . \ .$ Here α_1, α_2 are suitable complex numbers. Let us conjecture that the energy of the wave represented by Ψ is: $|\alpha_1|^2 e_1 + |\alpha_2|^2 e_2 + \ . \ . \ .$ Then the energy is conserved in the Schroedinger dynamics. We should not expect energy to be conserved in the von Neumann dynamics. For that does not concern isolated systems.

7

Indeterminate Particles

In chapter 4 I rejected a semi-Kantian speculation in which states were described in terms of propensities to produce various results when observations were made. I promised a successor. This was to be a speculation in which particles had indeterminate position, and, perhaps, indeterminate momentum. In this chapter I develop this Indeterminate Particle Speculation. Of the three speculations which I consider tenable this is the closest to the Copenhagen School.

I AN OUTLINE OF THE SPECULATION

As a preliminary, I shall try to remove an obstacle in the way of accepting the thesis that an electron, say, has an indeterminate location. It is the common sense objection that 'we all know' that an object cannot have an indeterminate location. I ask: Do 'we all know' this because of a priori insight or simply because of our experience? I submit it would have to be the latter. So what does experience tell us in this respect? Not that there is no indeterminacy but that there is too little to be noticed with the naked eye or even microscopes. If common sense were right Planck's constant would be zero. And common sense is not far wrong. Planck's constant is small. So the indeterminacy is not noticeable, which is why we do not notice it. Here there is an analogy with smooth surfaces. Common sense tells us that most glass objects have no surface irregularities. And common sense is not far wrong. The irregularities are small.

In the formalism of quantum theory we assign 'probabilities' to various regions of space. But as far as this formalism is concerned these 'probabilities' need not be probabilities. All that is required is that we assign to each region of V a number $\mu(V)$, usually taken to be between 0 and 1, such that for two regions U and V:

$$\mu(U \cup V) + \mu(U \cap V) = \mu(U) + \mu(V).$$

I propose treating such 'probability' assignments as indeterminate locations. The special case of a *determinate* location at q is given by a 'probability' assignment μ such that $\mu(V) = 1$ if q is in V, $\mu(V) = 0$ otherwise. If that is not so, but $\mu(V) = 1$ for some small region V around q then the indeterminate location *approximates* q.

I begin by describing the version of this speculation which I consider to be the simplest. I call it the Indeterminate Location Speculation. I consider other versions in section IV below.

The Indeterminate Location Speculation in some ways resembles the Determinate Particle Speculation, developed in chapter V. However, instead of a probability distribution for determinate locations we have a 'probability' distribution corresponding to an indeterminate location. Consider a single particle. At a given time, we have a 'probability' assignment μ for location. We may consider the rate of change of the 'probability' of location, but it does not follow that there is a 'probability' for the rate of change of location, that is, for velocity. Now on the other versions of the Indeterminate Particle Speculation there *is* a 'probability' distribution for momentum and hence velocity. But on this version no 'probability' distribution for momentum is posited.

To illustrate how we can fail to have 'probabilities' for velocity, consider the indeterminate location of a particle which is somewhere on a circle but at no particular point on that circle. Suppose its 'probability' distribution is, at all times, uniform. That is, suppose any point on the circle is as 'probable' as any other. Then the indeterminate location is constant, so its rate of change is zero. However this constant 'probability' distribution for location is compatible with *any* distribution for (angular) velocity, provided the velocity is independent of the location on the circle. Hence it does not imply a distribution for velocity. On the Indeterminate Location Speculation the only indeterminate properties are location and spin.

Not every non-literal interpretation of the 'probabilities' will give us a version of the Indeterminate Particle Speculation. For consider a fluid, or any other extended massive object of non-uniform density. Let the mass contained in a region U be $\mu(U)$. Then μ has all the formal characteristics of a probability measure. Yet we do not have a particle of indeterminate location, but rather an extended object of non-uniform density.

Because both a particle of indeterminate location and an extended object (or fluid) of variable density can be represented by a formal 'probability' measure μ, we have several useful analogies. Thus the extended object of non-uniform density has a *centre of mass*, the point about which it balances in any uniform gravitational field. Likewise, we can consider the 'mean' location of a particle of indeterminate location. This is the point which would be the centre of mass if the 'probability' distribution represented the non-uniform density of an extended object. I call this the *associated determinate location*. In the classical approximation, where indeterminacy is ignored, this associated determinate location may be taken as the particle's location. The particle's mass times the rate of change of the associated determinate location I call the *associated determinate momentum*. The associated determinate location and momentum behave just like their counterparts in the Wave Speculation. By Ehrenfest's theorem (Messiah, 1961, p. 218), for instance, the associated determinate location and momentum satisfy the classical equations of motion.

Consider the analogy with a fluid again. We can define a mass density for an object of non-uniform density. Likewise, we have a 'probability' density for the location of a particle. For a small region V, the 'probability' of a particle being in V approximates the density (at any point in V) times the volume of V.

A possible reason for not positing an indeterminate velocity is that the only natural way to posit one, namely the Indeterminate History Speculation (to be discussed in section IV), would lead to a joint 'probability' distribution for position and momentum. And we might want to avoid the negative 'probabilities' which would result. Here it suffices to note that no such reason prevents us considering joint 'probability' distributions for location and spin. We may consider all pairs of locations and directions of spin. Any 'probability' distribution over these will be interpreted as an indeterminate location-cum-spin. Since the distributions for spin and for location are not independent, we cannot treat the indeterminate location-cum-spin as the conjunction of an indeterminate location and an indeterminate spin. In the next section, when I have provided an account of indeterminate properties, I shall show that there is no difficulty with having an indeterminate location-cum-spin rather than an indeterminate location and an indeterminate spin.

Guided yet again by the analogy with a fluid, we can introduce the *current* $J(q)$ at q. This will be given by the same formula as the propensity current in chapter 5 or the current discussed in chapter 6.[1] And, as before, we may take the quantum state to specify and be specified by the 'probability' density together with the current.

In neither of the previous two chapters did we assign any fundamental role to the current. But in the Indeterminate Location Speculation we shall. The current plays a role analogous to dispositional momentum in classical mechanics. It is appropriate, therefore, to say more about it. As a first approximation, we may say that the current at q specifies the velocity the particle *would have* if it were to be observed to be at q. But this is not qutie right. For we do not have YNOs for determinate location. Instead the following holds:

> Suppose a YNO is made for whether the particle is in some small region V around q. And suppose there is a Yes result. Then the current at q (even prior to the observation) approximates the associated determinate momentum of the particle after the observation.[2]

Thus on the Indeterminate Location Speculation, we have no 'probability' distribution for velocities. Instead we have (to any desired approximation) a *potential* velocity for each point in space q, namely the velocity the particle would have were it observed to be near q. This gives us some idea of what the current is. But it leaves us with a problem. Why should something as fundamental as the current be describable in terms of what *would* happen if various YNOs were made? In section V, I shall show how we can redescribe the current in a non-dispositional fashion.

I now turn to systems of several indeterminate particles. First, this description is inaccurate. They should be described as indeterminate systems of several particles. For just as we require an indeterminate location-cum-spin rather than an indeterminate location and an indeterminate spin, we also require an indeterminate n-tuple of locations for n-particles, which is not just the conjunction of n indeterminate locations, one for each particle. The reason for this is that the 'probability' distributions for the particles need not be independent of each other.

We shall have a current, analogous to that considered on the Wave Speculation. For an n-particle system it has 3n components and is assigned to each n-tuple of locations.

The von Neumann dynamics for the measurement of locations is straightforward. An eigenstate for the YNO of whether the particle is in V is a state with an indeterminate location which is (determinately) contained in V, that is one for which $\mu(V) = 1$. And in the von Neumann dynamics, when the YNO is observed, there is a propensity for a state to become an eigenstate. This propensity equals $\mu(V)$. Here it is tempting to say that this tells us what the indeterminate location is: it is the propensity to become an eigenstate for various position YNOs. This is true but misleading. It is true because the indeterminate location can be described as such a propensity. It is misleading because it suggests the semi-Kantian speculation, according to which the best we can do is to describe it this way. On the Indeterminate Particle Speculation the location-property of the particle may be described non-dispositionally, as in the next section.

On the Indeterminate Location Speculation the expected value of the observation of location turns out to be the associated determinate location. So Mean Veridicality is satisfied for location. Likewise the expected value for the observation of momentum turns out to be the associated determinate momentum.[3] Hence if we take the associated determinate momentum to be the quantity we are aiming to observe when we observe momentum then we have Mean Veridicality for the observation of momentum.

Related to indeterminacy is the question of bivalence. Suppose a particle has an indeterminate location given by an assignment of 'probabilities' μ. And suppose V is a region in space. What truth-value does 'The particle is in V' have? I begin with a word of caution. The truth-value of a sentence partially depends on how it is standardly used. But we do not have *standard* usages governing predicates in situations where the properties are indeterminate. For our language has evolved against a background in which all properties were taken to be determinate. Hence any discussion of this topic can only be by way of saying what is a *natural* extension of the usage we do have. In this spirit I propose that indeterminacy be handled like equivocity. Indeed the usual sorts of equivocity could be called equivocity *de dicto* and indeterminacy could be called equivocity *de re*.

Why should we handle indeterminacy like equivocity rather than vagueness, which is the other obvious analogue? In answer to this, I first note that since we are only *likening* indeterminacy to equivocity or to vagueness, the differences between equivocity and vagueness are not especially significant. None the less, I prefer to liken indeterminacy to equivocity because of the determinateness of the boundary of indeterminacy. For instance, if a particle is indeterminately somewhere in region V, the region V is itself specified in a quite determinate fashion. This is unlike the situation that occurs with vagueness. Consider the vague predicate 'bald'. It is not as if 'definitely bald' and 'definitely not bald' were themselves precise. No, even the boundary around the definitely bald is itself somewhat vague. By contrast, in paradigm cases of equivocation, we equivocate between what is itself not equivocal.

Here is a typical piece of equivocity de dicto: 'She checked the child's progress.' That could mean 'She stopped the child from progressing for a while' or 'She assessed how well the child was progressing'. Suppose a boy in an art class is a bit nervous about his piece of sculpture. An art teacher might set about assessing how well he is doing and so intentionally check his progress in the latter sense, but unintentionally check his progress in the former sense. Then, in spite of the equivocity, the statement 'She checked his progress' is, I take it, true. If he was progressing well and she ignored him, in spite of the equivocity it is false. However, if he continued his work while she inspected it, then in one sense it would be true, in another false. In that case we could say that 'She checked his progress' is both true and false, neither true nor false, or half true and half false. To decide which to say, I first define *pure truth* as truth without any admixture of falsity. I then stipulate that when I say 'true' hereafter I mean 'purely true'. And similarly with falsehood. So in this, and similar cases, the statement is neither true nor false. It might be objected to my use of this example that we should consider how the speaker intended the statement to be taken, or how the audience took it. But the speaker could be deliberately punning and the audience could recognize this. Or both speaker and audience could be too confused to recognize the possibility of equivocation. So the example is not undermined.

To apply these rules to equivocation de re, that is indeterminacy, we need to say what determinate locations the indeterminate one is equivocal between. Now the formalism will provide us with a 'probability' density.[4] We may say that the indeterminate location is equivocal de re between all points for which this density is non-zero. Let that region be R. Then we have:

(i) 'The particle is in U' is true if $R \subseteq U$.
(ii) 'The particle is in U' is false if $R \cap U = \emptyset$.
(iii) Otherwise 'The particle is in U' is neither true nor false.

This failure of bivalence has not arisen from the identification of truth with warranted assertibility, but by a natural extension of the rules for the assignment of truth from the familiar equivocation de dicto to the unfamiliar equivocation de re. None the less they still indicate a deflationary attitude towards the location of particles. We accept the statement that the particle is in R, and we say it is true. But it is not as if the particle has a determinate location which is in R. So our attitude to location is a somewhat cautious or deflationary one.

II BUT WHAT IS AN INDETERMINATE LOCATION?

I have stated a variant of the Indeterminate Particle Speculation. In it I consider assignments of 'probabilities'. But what are these 'probabilities'? In this section I shall critically examine various suggestions. Regrettably, I am both the prosecution and the defence. For, as far as I know, no one has previously discussed what an indeterminate location might be. So I have had to think of some proposals myself.

The first proposal is that an indeterminate location is simply a region of non-zero but finite volume. The particle is somewhere in that region but there is no one position at which the particle definitely is. One way to make this precise is to deny bivalence for location, and to say (as in the previous section) that for a particle with indeterminate location R we have:

(i) 'The particle is in U' is true just in case $R \subseteq U$.
(ii) 'The particle is in U' is false just in case $R \cap U = \varnothing$.
(iii) Otherwise 'The particle is in U' is neither true nor false.

The obvious objection to this proposal as an account of indeterminate location is that we require an assignment of 'probabilities'. So the proposal is incomplete until we are told how the 'probabilities' are specified by the region R. If U is any region we could assign 'probability' μ_R where $\mu_R = K \times \mathrm{vol}(U \cap R)$, for some constant K. This satisfies the requirement that:

$$\mu_R(U \cup V) + \mu_R(U \cap V) = \mu_R(U) + \mu_R(V).$$

However, the assignments of 'probability' given by the formalism of quantum theory are highly varied. And we would require that for each such assignment μ, there is a region R such that μ_R approximates μ. Can this be done? There is some flexibility here, for we have not specified in what way μ_R approximates μ. Here is my suggestion. Given any δ, however small, any distance ϵ, however small, and an assignment of 'probabilities' μ, which arises in quantum theory, we can find a region R such that for any balls B of radius greater than ϵ, the difference between $\mu_R(B)$ and $\mu(B)$ is less than δ times the volume of B.[5] This seems a suitable notion of approximation. But the proposal is still not entirely satisfactory. For the region R will be a highly intricate one. In two dimensions we might have something like figure 7.1. Here the grid consists of lines much closer together than ϵ. The shaded areas are parts of R, whose total area is 1. The proportion of each square occupied by the shaded region R will vary. It is greater where the 'probability-density' is greater. The intricacy of this region is a possible objection to the proposal. So while not entirely dismissing it, I shall provide an alternative.

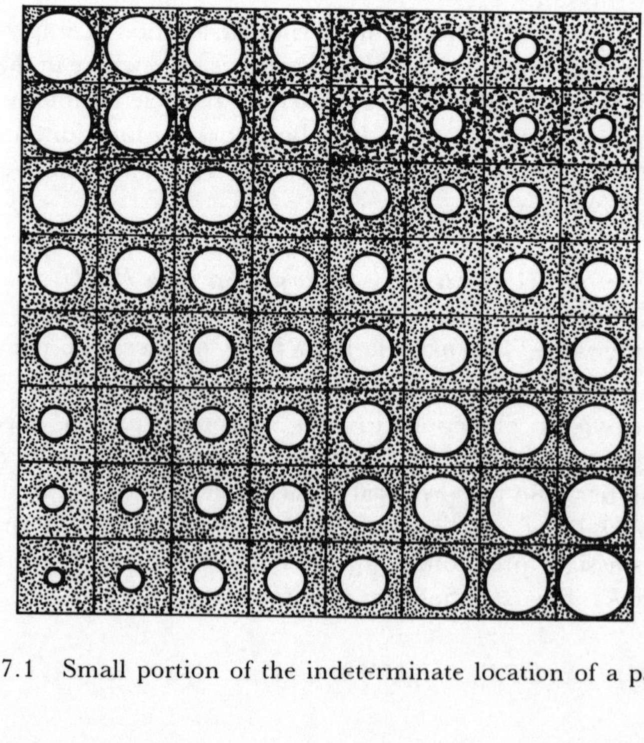

Figure 7.1 Small portion of the indeterminate location of a particle.

A rather different proposal is to introduce *degrees* of truth. Not only do we have falsity (0) and truth (1) but we have degrees of truth measured by numbers from 0 to 1. This has been proposed in the context of vagueness (Priest, 1979b), and would also be natural if we were to identify truth with warranted assertibility. For the latter does admit of degrees. Thus for the particles with an indeterminate location given by μ, we would assign the degree of truth $\mu(V)$ to the statement 'The particle is in V'.

The difficulty with this proposal is that degrees of truth are familiar only in the contexts of vagueness, truth as warranted assertibility, and degrees of approximation to truth. And none of these seems applicable to the case of quantum theory. I grant that indeterminacy is somewhat *like* vagueness, although I consider it more like equivocity. But if it were *literally* a matter of vagueness then greater precision should remove it. We should then provide this precision, and so no longer need the Indeterminate Particle Speculation. Nor is truth here a matter of warranted assertibility. As I argued in chapter 3, it would have to be what was assertible by someone who knew quantum theory, which is circular. Nor can we appeal to approximation to the truth. Perhaps the whole of quantum theory is only an approximation to the truth. But that is not the issue. If we said that 'The particle is in U' is just an approximation to the truth, yet 'The particle is in V' is true, then we should simply ignore the former statement in favour of the latter. We do not need approximations to the truth if we also have truths as answers to our questions.

Therefore if we introduce degrees of truth we are introducing a novel kind of degrees of truth. The only justification for this would be that, given our language, the best way we have for describing the physical world is to say that various sentences are true in varying degrees. This would be an indirect way of describing the world, one which operated by means of a detour through considerations of truth. It is preferable, where possible, to describe the world more directly. Hence degrees of truth should be the *last resort* of the theorist giving an account of indeterminate properties. Because I shall provide an alternative, more direct, description I shall not rely on degrees of truth.

Another proposal is to take the indeterminate locations as properties in some way analogous to determinate locations, but not analysable in terms of them. This proposal would be given content by claiming that there is an indeterminate location corresponding to each 'probability' distribution. This proposal would be satisfactory if we had an independent understanding of what the 'probability' in question is. As it is, it is not helpful to be told that we have something *like* a location which corresponds to something satisfying the purely *formal* characteristics of probabilitics.

Theft has failed. We are driven to honest toil. What we need is a systematic theory of indeterminate locations, according to which they are all analysable in terms of determinate properties using a few fundamental operations. I now provide such a theory. First I propose that we consider a primitive operation (which I call *indeterminate disjunction*) of a finite number of pairwise incompatible properties. I write the indeterminate disjunction of the properties $F_1, \ldots, F_n$ as $F_1 \iota \ldots \iota F_n$. To get a grip on this new concept, we should consider some truth and falsity conditions. This requires a connection between properties and predicates. It is rather like ordinary implication, but it crosses from one category, that of properties, to another, that of certain linguistic items, namely predicates. I shall say that the property F *π-implies* the predicate 'H' whenever, of necessity, if '*a*' designates an item with F, then '*a* is H' is true. And I shall say that the property F is *π-inconsistent* with the predicate 'H' whenever, of necessity, if '*a*' designates an item with F, then '*a* is H' is false.

Suppose Q is the property of being at a determinate location corresponding to the predicate 'is at q', and V is a region in space. Then Q π-implies 'is in V' if q is at point in V; and Q is π-inconsistent with 'is in V' otherwise.

I assume that indeterminate disjunction is commutative, that is, the order of the disjuncts does not matter. I shall not, however, assume associativity. In addition we have three rules:

(i) $F_1 \iota \ldots \iota F_n$ π-implies 'H' just in case all of $F_1, \ldots, F_n$ π-imply 'H'.
(ii) $F_1 \iota \ldots \iota F_n$ is π-inconsistent with 'H' just in case all of $F_1, \ldots, F_n$ are π-inconsistent with 'H'.
(iii) Assuming, as I do, that the F_i are pairwise incompatible, $F_1 \iota \ldots \iota F_n$ is π-inconsistent with the predicate 'has F_i'.

Rules (i) and (ii) are those for ordinary disjunction. But rule (iii) clearly distinguishes the case of indeterminate disjunction.

Using these rules, if Q_i is the property of being at q_i, if V is a region in space, and if a has the indeterminate location-property $Q_1 \iota \ldots \iota Q_n$, then we find, as required, that:

(a) '*a* is in *V*' is true just in case all the q_i are in *V*.
(b) '*a* is in *V*' is false just in case none of the q_i are in *V*'.
(c) Otherwise '*a* is in *V*' is neither true nor false.

The indeterminate locations obtained in this way I call *finitely* indeterminate locations. Generalizing, if we form the indeterminate disjunction of any finite number of determinate properties, we have *finitely indeterminate* properties.

Assuming that *F* and *G* are incompatible, it follows that $F \iota G$ is incompatible with both *F* and *G*. (And in general, $F_1 \iota \ldots \iota F_n$ is incompatible with all the F_i.) That is because, by rule (iii), $F \iota G$ is π-inconsistent with 'has *F*', while *F* π-implies 'has *F*'. Now, on the Indeterminate Particle Speculation, spin properties are always indeterminate. That is, the spin is always *somewhere* in the up hemisphere with respect to a given direction without being in any one determinate direction. And the same holds for location, for particles never have a perfectly determinate location. Therefore there will be instances of indeterminate disjunctions whose disjuncts are never instantiated. Provided the operation of indeterminate disjunction is considered a genuine operation on properties realistically conceived, this commits one to the thesis often called Platonism, namely that there are uninstantiated properties. The nominalist, of course, will apply whatever strategy he usually applies to prima facie commitments to properties. So this will not worry him. But it will worry those such as Armstrong (1978, vol. I, p. 113), who insists that all properties are instantiated. For such thinkers uninstantiated properties will be a cost - - perhaps too high a cost – of the Indeterminate Particle Speculation (and of the Wave Speculation).

Another question that should be discussed before we proceed is whether indeterminate disjunction is associative. If it is, then we may treat the indeterminate disjunction of any finite number of disjuncts as the result of successive operations of the indeterminate disjunction of just two properties. This would be a simplification. But there is a reason for not assuming associativity, which I now state. It would unnecessarily complicate the speculation if the indeterminate properties did not determine the 'probabilities'. Let us assume they do. Then $F \iota G$ is the property of being *F* with probability 1/2 and *G* with probability 1/2. To make this more perspicuous let us write $F \iota G$ as $1/2\,F + 1/2\,G$. Likewise $F \iota G \iota H$ is to be written $1/3\,F + 1/3\,G + 1/3\,H$.

But now we find that:

(a) $F \iota G \iota H = 1/3F + 1/3G + 1/3H$
(b) $(F \iota G) \iota H = 1/4F + 1/4G + 1/2H$
(c) $F \iota (G \iota H) = 1/2F + 1/4G + 1/4H$.

Hence we do not have associativity.

I have already mentioned the correlation of 'probability' distributions with indeterminate disjunctions. This is straightforward. We assign to each disjunct the same 'probability'. Consider, for example, the indeterminate disjunction of the n determinate locations: being at q_1; being at q_2; . . .; being at q_n. The 'probability' distribution is one for which all the n positions $q_1, \ldots, q_n$ have equal 'probability'. So if a region V contains just m of the n positions it is assigned a 'probability' of m/n.

We may approximate any 'probability' distribution by one derived from a large but finite indeterminate disjunction of determinate locations. The greater the 'probability' density, the closer the points have to be. Consider for example a 'probability' distribution in one dimension given by the curve in figure 7.2.

It is zero until near $x = 0$, then it rises to 1/2 where it stays constant until $x = 1$, after which it gradually declines. We could approximate this by means of an indeterminate disjunction of the points of figure 7.3. There are few points until near $x = 0$, then they are uniformly spaced until $x = 1$ after which the spacing becomes greater. The mathematical details of this approximation are straightforward, but tedious, and I shall not go into them.

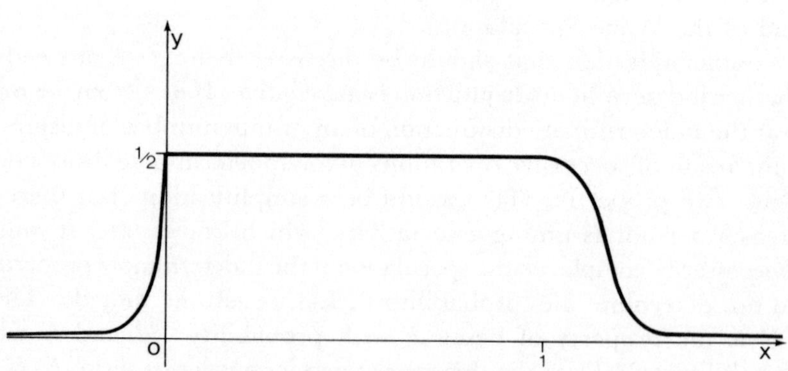

Figure 7.2 A 'probability' density function.

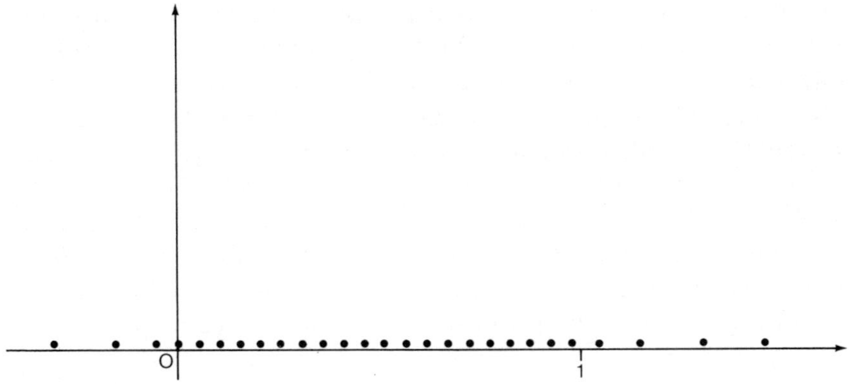

Figure 7.3 Indeterminate location corresponding to the 'probability' density of figure 7.2.

Every 'probability' distribution, then, is approximated by one of those correlated with a finitely indeterminate location. This provides two ways of further developing the speculation. The first is to insist that it is quantum theory which is only approximate, and that, in fact all indeterminate properties are finitely indeterminate.[6] However this will prevent certain physically possible processes from being continuous. For example, consider an electron which is indeterminately in Athens or in Jerusalem. The quantum theory allows its state to change in such a way that the 'probability' of its being in Athens increases continuously. Yet if all indeterminate properties are finitely indeterminate, there would have to be discontinuity.

That these processes be continuous is a desideratum for a speculation, but no more than that. So we are entitled to reject continuity and to insist that all properties are *finitely* indeterminate. However it is possible to preserve continuity, at a cost. The cost is that we have to allow *infinitely* indeterminate properties which themselves are approximated by, and indeed are limits of, finitely indeterminate properties. The appropriate concept of approximation is that the associated 'probability' distributions approximate each other.[7] All indeterminate properties and, in particular, all indeterminate locations are then analysable into determinate locations using two operations, that of finite indeterminate disjunction and that of taking the limit.

We are now in a position to show that there is nothing peculiar about properties obtained by first conjoining determinate properties and then taking indeterminate disjunctions and limits. A theory of properties and relations will include an account of various operations, of which the simplest is conjunction. I have proposed one or two new operations, indeterminate disjunction and perhaps the limit operation. It would be a further complication of the theory to put extra constraints on the order in which these operations are performed. To refrain from so doing is therefore an advantage not a disadvantage. Hence there is nothing peculiar about first conjoining, then taking finite indeterminate disjunctions, and finally taking limits. In particular there is nothing peculiar about indeterminate location-cum-spin.

III ACTION-AT-A-DISTANCE AND INDETERMINATE LOCATION

I have denied that action-at-a-distance is a difficulty for any speculation about quantum theory. None the less it is interesting to discover what happens to prima facie cases of action-at-a-distance on the Indeterminate Particle Speculation. A first thought here might be that a particle of indeterminate location is all over the place, like a wave, and so there is no distance between it and what affects it. But that is too swift. An indeterminate location is not the location of something extended in space, like a wave or a fluid. What is going on is that the failure of bivalence carries over to action-at-a-distance.

We say that there is action-at-a-distance if some factor directly influences a system or subsystem separated from it by a positive distance. For this to be *true*, the factors would have to be distant from the region R where the 'probability' density is non-zero. And this never occurs. In the twin-slit experiment this region, at the time when the electron reaches the slits, comprises both slits. Likewise in the situation described in Bells' Inequality the two-particle system being affected by the observation has an indeterminate joint location which embraces the location of both observations of spin. We might protest that we are interested in the locations of the individual particles, not in the joint location. But there are no such individual locations, determinate or otherwise.

However, the statement that there is action-at-a-distance is not *false* either. For it is not false that the particle goes through one of the slits and is influenced by the other. On the Indeterminate Particle Speculation then, that there is action-at-a-distance is neither true nor false. I leave it to those who think of action-at-a-distance as a difficulty to decide whether this failure of bivalence removes the difficulty or not.

IV VARIANTS

On the Indeterminate Location Speculation, the only indeterminate properties were location and spin. I now consider two variants in which we allow other indeterminate properties. According to the first, which I call the Indeterminate Observable Speculation, there are indeterminate properties corresponding to all and only the observables.

First, consider momentum. We cannot define an indeterminate velocity in terms of the indeterminate locations at various times. But that does not prevent us positing a dispositional momentum, that is, a property which can be described as a tendency to change location. Suppose a particle of mass m has a determinate dispositional momentum mv. Then its perhaps indeterminate location will tend to change accordingly. For instance, suppose the indeterminate location changes from $Q_1 \iota \ldots \iota Q_n$ to $Q_1^* \iota \ldots \iota Q_n^*$ in time Δt. Then if Δt is small, the vectors from Q_j to Q_j^* should approximate $v\Delta t$.

We can define a determinate kinetic energy in terms of the determinate momentum; a determinate angular momentum in terms of determinate locations and determinate momentum; and so on. In this way, we can posit all the determinate properties corresponding to the observables. On the Indeterminate Particle Speculation, these determinate properties are never instantiated.[8] However, on the variant being considered, for each observable, whether it is location, momentum, energy, angular momentum, or something more complicated, there is instantiated an *indeterminate property* which is the limit (if necessary) of finite indeterminate disjunctions of the determinate properties.

Notice that a particle need not have an indeterminate position-cum-momentum. For simplicity, suppose the location is $Q_1 \iota Q_2$ and the momentum is $P_1 \iota P_2$. Then the particle has a conjunctive property $(Q_1 \iota Q_2)$ & $(P_1 \iota P_2)$. But I have not, and shall not, posit the distributivity rule which would enable us to infer that the particle had the indeterminate position-cum-momentum $(Q_1 \& P_1) \iota (Q_1 \& P_2) \iota (Q_2 \& P_1) \iota (Q_2 \& P_2)$. As a consequence, on the Indeterminate Observable Speculation, we are not threatened with a joint distribution for position and momentum.

By contrast, there is a further variant which *does* provide such a joint distribution. It is the Indeterminate History Speculation. First, let me explain what I mean by a *determinate* history. For a single particle, it is the assignment of a location (and if necessary a determinate spin) to each time *t*. Thus a classical particle would have a determinate history. On the Indeterminate History Speculation a particle has an *indeterminate* history, namely the indeterminate disjunction of determinate histories. (Or perhaps their limit.) For a system of several particles, a determinate history is the assignment of the locations and spins of the various particles to each time *t*.[9] The indeterminate history is the indeterminate disjunction of these. The indeterminate history specifies the quantum state.

The Indeterminate History Speculation may be compared to Feynman's theory of particle interactions (Feynman, 1948, 1949). On Feynman's theory, we consider probability distributions over what I call histories of systems of many particles. Then the Schroedinger dynamics is specified by a certain constraint on the probability distribution.[10] There may be mathematical problems in handling probability distributions over histories.[11] And this may perhaps be a difficulty with Feynman's theory as a piece of detailed theoretical physics. But there is no incoherence in the concept of such a probability distributions. So this difficulty does not apply to our metaphysical speculation.

It is easy enough to characterize 'probability' distributions for location, for momentum and, indeed, for location-cum-momentum, in terms of that for the histories.[12]

This is a convenient place to discuss the Converse of Weak Veridicality. This holds for location and spin on the Indeterminate Location Variant and for all observables on the other variants. As on the Wave Speculation, the Converse of Weak Veridicality does not hold for momentum on the Indeterminate Location Variant. For on that variant, the quantity to be observed would have to be taken to be the associated determinate momentum. This is always a precise value. Yet the quantum state will not be an eigenstate for a precise momentum.

This is also a convenient place to discuss the Conservation of Energy. Now we could *associate* a determinate energy with an indeterminate particle, in the same way as we associated an energy with a wave. In that case the energy would be conserved. On either the Indeterminate Observable Speculation or the Indeterminate

History Speculation, however, we may also consider the indeterminate energy. In that case, the Conservation of Energy is like action-at-a-distance. Because the energy is indeterminate among a range of values, which range is not altered in the Schroedinger dynamics, it is neither true nor false that energy is conserved. In this respect, then, the Indeterminate Particle Speculation could be considered indeterminate between the Wave Speculation and the Determinate Particle Speculation.

V THE PROBLEM OF THE CURRENT

In the Indeterminate Location Speculation, the current played a fundamental role, analogous to that of momentum in classical mechanics. Yet the current at q could only be characterized as the associated determinate momentum which the particle *would have if the particle were observed to be at* q. That such a fundamental quantity can only be described in this roundabout way is a disadvantage of that variant. In this respect the Indeterminate Observable Speculation is no improvement. For we cannot characterize the current in terms of the separate 'probability' distributions for position and for momentum. We would require a joint distribution, which we do not have on that variant.

On the Indeterminate History Speculation the current has a straightforward interpretation. That is because we *do* have a joint distribution for location and momentum defined in terms of the indeterminate history. So for a given location, we have a 'probability' distribution for momentum. It specifies a 'mean' momentum which is just the current at q. In this way, the current is characterized in terms of the indeterminate history.

Unfortunately, as I noted in chapter 2, there is reason to expect any joint 'probability' distribution to have negative values. Hence, there must be a negative 'probability' for some histories of a particle.[13] On the Determinate Particle Speculation, we got round this difficulty by proposing negative propensities. But these would be quite mysterious *outside* of the context of large collections of systems. So we should not appeal to negative propensities here.

In the next section I provide a way of handling the negative 'probabilities'. Provided it is acceptable, we should prefer the Indeterminate History Speculation because of the way it explains what the current is.

VI NEGATIVE MASS

We have positive and negative charge, we can have equal and opposite forces. Why not negative mass?[14] We may well be suspicious about a metaphysician proposing new kinds of particle. That is the job of physicists. But I am making no such proposal. All I am proposing here is that we consider the *uninstantiated* property of having negative mass in order to speculate about actual particles whose mass is positive.[15]

Given this hypothesis, it is easy to account for the negative 'probabilities'. For simplicity concentrate on a single particle whose mass is usually taken as $+m$. And consider a determinate history h. Then we have the determinate property H^+ of its mass being $+m$, and its history being h. We also have a determinate property H^- of its mass being $-m$, and its history being h. Neither of them will be instantiated. But we may form an indeterminate disjunction of such positive and negative histories, $H_1{}^+ \iota \ldots \iota H_m{}^+ \iota H_{m+1}{}^- \iota \ldots \iota H_{m+n}{}^-$ (here, I assume that $m > n$). And we could take limits of these. The 'probability' distribution μ associated with this indeterminate history can be described as follows. For any set X of determinate histories, let $\mathrm{Card}^+(X)$ be the cardinality of $\{h_1, \ldots, h_m\} \cap X$. And let $\mathrm{Card}^-(X)$ be the cardinality of $\{h_{m+1}, \ldots, h_{m+n}\} \cap X$. Then:

$$\mu(X) = (\mathrm{Card}^+(X) - \mathrm{Card}^-(X))/(m - n).$$

This is the natural generalization of the 'probability' assigned to such indeterminate histories as $H_1{}^+ \iota \ldots \iota H_m{}^+$, but it takes negative as well as positive values.

By appealing to negative-mass particles, then, we can handle the negative 'probabilities'.

VII OBSERVATION AND THE INCREASE OF DETERMINACY

If we take seriously the idea that a particle can have a location without having any determinate location, then it is natural to conjecture that observation increases determinacy. In this section I consider whether this conjecture can be defended.

Let us first consider the case of the location of a single particle. And suppose that *before* we make a location YNO, it has the indeterminate location property $Q_1 \iota \ldots \iota Q_n$. Here Q_k is the property of being at q_k, for $k = 1, \ldots, n$. Suppose the YNO is whether the particle is in region V. And finally suppose that $q_1, \ldots, q_m$ are in V but $q_{m+1}, \ldots q_n$ are not. Then the von Neumann dynamics results in a probability m/n of a Yes result. And if there is a Yes result, the 'probability' distribution for location after the observation is that which would be given by the indeterminate location $Q_1 \iota \ldots \iota Q_m$. Admittedly, it would also be given (approximately) by other indeterminate locations, such as $Q_1 \iota \ldots \iota Q_m \iota Q_1^* \iota \ldots \iota Q_m^*$, where Q_k^* is the property of being at q_k^*, a point very near q_k. But by far the simplest proposal is that after the observation, the location property is $Q_1 \iota \ldots \iota Q_m$.

There are grounds, then, for asserting that the observation of location typically results in an increase in the determinacy of location. We pass from indeterminacy between the set of locations $\{q_1, \ldots q_n\}$ to indeterminacy between $\{q_1, \ldots, q_m\} = \{q_1, \ldots, q_n\} \cap V$.

The case of an infinitely indeterminate location is not so straightforward. For there is no obvious sense in which one limit of finite indeterminate disjunctions is more determinate than another. Perhaps the simplest account of what it is for an infinitely indeterminate location to be more determinate than another is to consider the associated regions where the 'probability' density is non-zero. The smaller the region the more determinate the location.[16] If we adopt this proposal, then, once again, observing location typically results in an increase in determinacy. In both cases an exception occurs when the state is an eigenstate even before observation. In that case there is neither an increase nor a decrease in determinacy.

On the Indeterminate Observable Speculation we may extend this to all observations. Any observation is an increase in determinacy of the quantity being observed unless the system is already in an eigenstate.

Notice that I have not defined an increase of determinacy in terms of the variance of the observable. As I have characterized it, the determinacy can increase even if the variance – the measure of spread – also increases. For example, consider an indeterminate location $Q_1 \iota Q_2 \iota Q_3$, where the corresponding points in space, q_1, q_2 and q_3 are on a straight line, with q_2 between q_1 and q_3. Suppose q_2 is in V but neither q_1 nor q_3 are. Then a No result to the YNO of whether the particle is in V will result in an increase in the variance of the associated 'probability' measure. Yet it also results in an increase in determinacy. It follows that Heisenberg's Uncertainty Principle does not show that an increase in determinacy in location, as a result of the observation of location, is accompanied by a decrease in the determinacy of momentum.

The conjecture that observation typically increases determinacy will apply to the Indeterminate History Speculation as well. Here it takes an especially attractive form. A Yes result for any YNO restricts the set of determinate histories between which there is indeterminacy to the set of those for which the Yes result is veridical.

VIII ASSESSING THE INDETERMINATE PARTICLE SPECULATION

The advantages and disadvantages of the Indeterminate Particle Speculation are almost entirely the same as those for the Wave Speculation. We require a novel operation, indeterminate disjunction. Probably we also have to consider limits of indeterminate disjunctions. As a consequence, we are committed to uninstantiated properties, which might be thought a disadvantage. But for the same reason, we are spared irreducible relations of high polyadicity, which propensities turn out to be on the Determinate Particle Speculation. And we do not need the complicated laws of nature required to handle those propensities.

The Indeterminate Particle Speculation is certainly no worse than the Wave Speculation when it comes to Schroedinger's Cat and the Problem of the Two Dynamics.

There is a respect in which the Indeterminate Particle Speculation is not as satisfactory as the Wave Speculation. It fails to give us quite as intuitive an account of what is involved in superposition. This difficulty should not be overemphasized, however. For we can characterize the superposition of two states as a third state which is 'between' them, as in chapter 2. Another minor disadvantage is that, unlike the Wave Speculation, the Indeterminate Particle Speculation does not result in the truth of the classical Conservation of Energy.

NOTES

1 See note 19 to chapter 5.
2 Strictly it is the limit of the associate determinate momenta which would result, as we let the volume of the region around q tend to zero.
3 See section I of chapter 6 for a discussion of these results.
4 A probability density $f(q)$ is a function such that for any (measurable) region U, $\mu(U)$ is the integral over U of $f(q)$. Not every probability distribution will have a corresponding probability density. But any which are obtained from a wave function $\Phi(q)$ must. For then $f(q) = |\Phi(q)|^2$. Moreover if Φ satisfies Schroedinger's equation f must be continuous. And there can be no more than one continuous probability density function for a given probability measure.
5 The formalism gives us a 'probability' measure derivable from a density $f(q)$. Hence, given ϵ and δ, it is possible to find a region A and a constant K, such that $\mu(A) > 1 - \delta\epsilon^3$, and such that if $q \in A$ then $f(q) \leq K$. Then, for any region U, $\mu_A(U) \leq K$ vol (U). Now for any ball B of radius ϵ, $\mu(B) - \mu_A(B) \leq 1 - \mu(A) \leq \delta\epsilon^3 \leq 1/4\, \delta \operatorname{vol}(B)$. So it suffices to find a region R such that $|\mu_A(B) - \mu_R(B)| < 3/4\, \delta \operatorname{vol}(B)$ for any ball B of radius ϵ. To do this we divide the whole of space into countably many disjoint small cubes, C_i, of side η, where η is less than ϵ and less than $\delta\epsilon/16K\sqrt{3}$. In each C_i we select a region R_i such that $\operatorname{vol}(R_i) = \mu_A(C_i)/K$. Hence for any region U which is the union of some of the C_i we have:

(1) $\mu_R(U) = \mu_A(U)$

If B is a ball of radius ϵ, let D be the union of all the C_i which intersect B and let E be the union of all the C_i contained in B. Then

(2) $0 \leq \operatorname{vol}(D) - \operatorname{vol}(E) \leq ([1 + (\sqrt{3}\eta/\epsilon)]^3 - [1 - (\sqrt{3}\eta/\epsilon)]^3) \times \operatorname{vol}(B)$

Now $[1 + (\sqrt{3}\eta/\epsilon)]^3 - [(1 - (\sqrt{3}\eta/\epsilon)]^3 = 6\sqrt{3}\eta/\epsilon + 2(\sqrt{3}\eta/\epsilon)^3$
Since $\eta < \epsilon$, $(\sqrt{3}\eta/\epsilon)^3 \leq 3(\sqrt{3}\eta/\epsilon)$.
And $\eta \leq \delta\epsilon/16K\sqrt{3}$, so we have:

(3) $0 \leq \operatorname{vol}(D) - \operatorname{vol}(E) \leq (3\delta/4K) \times \operatorname{vol}(B)$

Now:

(4) $\mu_R(D) \geq \mu_R(B) \geq \mu_R(E)$

and:

(5) $\mu_R(D) = \mu_A(D) \geq \mu_A(B) \geq \mu_A(E) = \mu_R(E)$.

From (4) and (5):

(6) $|\mu_A(B) - \mu_R(B)| \leq \mu_R(D) - \mu_R(E) \leq K\,[\mathrm{vol}(D) - \mathrm{vol}(E)]$.

From (1), (3) and (6) we have:

(7) $|\mu_A(B) - \mu_R(B)| \leq 3/4\,\delta \times \mathrm{vol}(B)$,

as required.

6 The approximation in this case would be that the formalism tells us that the probability distributions are continuous, whereas in fact they would be discrete.

7 Suppose we are considering 'probability' measures μ and ν over a space S, for example over locations for a single particle. Then we may select a family F of suitable regions, say all balls in three-dimensional space. Then the least upper bound of $\{|\mu(U) - \nu(U)| : U \in F\}$ is a suitable quantity for measuring the degree to which the 'probability' measures approximate. Call this $\|\mu - \nu\|$. This enables us to define a Cauchy sequence of finitely indeterminate properties corresponding to 'probability measures' μ_1, μ_2 etc. as follows. Given any $\epsilon > 0$ there is an integer k such that if $m \geq k$ and $n \geq k$, $\|\mu_m - \mu_n\| < \epsilon$. We then *posit* infintely determinate properties which are *limits* of such Cauchy sequences and whose 'probability' measures are the limits of the sequence μ_1, μ_2, etc.

8 I have taken an eigenstate for a spin YNO to be a state in which the spin direction makes an acute angle with the up direction. No greater determinacy is possible for spin. For position and momentum we can, it seems, have states which approximate determinacy as well as we please. Yet in the standard formalism, there are no states of determinate position or determinate momentum. If we modify the formalism to allow for them, we shall still have uninstantiated determinate spin properties.

9 As I have stated it, I have not assumed any diachronic unity for the particles. If we do assume diachronic unity, then a determinate history for a system of such particles consists rather of several determinate histories of individual particles.

10 Let the histories be given by functions from time to space. Let the particle's mass be m. Then we have a 'probability' measure μ over a space S of such functions. From it is derived a joint distribution ν_t for location q and momentum p, where if W is any suitable set of pairs of location and momentum, we have:

$$\nu_t(W) = \mu(\{f \in S : \langle f(t), \mathrm{M}f\cdot(t)\rangle \in W\}).$$

This determines a wave function using Wigner's formula. Hence Schroedinger's equation puts a constraint on μ.

11 Because of the infinite-dimensional character of the 'space' of all possible histories.

12 See note 10 of this chapter.

13 If a set of position/momentum pairs is assigned a negative probability, the set of all histories which have position/momentum pairs in this set at a given time must have the same 'probability', which is therefore also negative.

14 Philosophers have an awkward habit of answering rhetorical questions. If by *mass* we mean a measure of the amount of stuff there is, then it cannot be negative. But if we just mean a quantity involved in various physical constraints, then it can. I take it the latter way.

15 See note 19 to chapter 2, for another reason for positing negative mass. Once again, we could ensure that there is no determinate negative mass.

16 More generally, any 'probability' measure μ will be the sum of a discrete and a continuous part. To the discrete part there corresponds a set of points D each of non-zero measure. The continuous part can be expressed as the integral of a 'probability density'. The 'probability density' obtained in this way is not unique. But any two will differ only on a set of zero (or infinitesimal) volume. So we can assign to each measure a set of points D together with an equivalence class C of regions differing by sets of volume zero (or infinitesimal). The measure μ_1 is then at least as determinate as measure μ_2 if D_1 is contained in D_2 and every member of C_1 is contained in some member of C_2. In this way the measures are quasi-ordered with respect to determinacy.

8

Final Speculations

I have presented three metaphysical speculations generated by quantum theory. The time has come for some conclusions. First, I bring together the comparisons I have made between those speculations. Then, mounting speculation on speculation, I discuss what conclusions we can draw about the physical world. Finally, I consider the connection between quantum theory and the metaphysics of persons.

I COMPARISONS

The best way to bring together the comparisons scattered throughout the last three chapters is to put them in tabular form and then comment on the resulting table. This I do below:

	Determinate Particle	*Wave*	*Indeterminate Particle*
The Two Dynamics			
(a) For materialists	No problem	Heroic metaphysics	Heroic metaphysics
(b) For dualists	No problem	Cartesian interactionism	Observation as increase in determinacy
Schroedinger's Cat	No problem	Peculiar	Failure of bivalence
(The Dyadic Programme)	Rejected	Rejected	Rejected
(Non-accidental correlations at a distance)	True	False	Neither
Veridicality	Weak Veridicality plus some Statistical Veridicality	Weak Veridicality plus some Mean Veridicality	Weak Veridicality plus some Mean Veridicality

	Determinate Particle	*Wave*	*Indeterminate Particle*
(Converse of Weak Veridicality)	Yes	For location, not momentum	For location, not momentum
Superposition	Not natural	Natural	Not natural
Mixture	Natural	Problem	Problem
Schroedinger dynamics	Complicated	Simple	Simple
Novel operations on properties	No	Yes	Yes
Conceptual innovations	Tied propensities	Polywaves	Indeterminate properties
Intrinsic occurrent descriptions of states	Not available	Available	Available
Relations of high polyadicity	Unanalysable	Analysable	Analysable
(Uninstantiated properties)	No new commitment	Commitment	Commitment
(Classical Conservation of Energy)	False unless negative energy posited	True	Neither true nor false, unless negative energy posited
Negative mass	Not required provided tied propensities are acceptable	Not required	Required on the Indeterminate History Variant
Successes not shared by both the rival speculations	Two Dynamics for materialists; Schroedinger's Cat mixture; no novel operations; (no new commitment to uninstantiated properties)	Superposition; (action-at-a-distance); (classical Conservation of Energy)	Two Dynamics for dualists
Difficulties	Schroedinger dynamics; no intrinsic occurrent description of state; unanalysable relations of high polyadicity	Two Dynamics for materialists, and to a lesser extent for dualists; Schroedinger's Cat; mixture; (uninstantiated properties)	Two Dynamics for materialists; (uninstantiated properties); negative mass on the Indeterminate History Variant

The Problem of the Two Dynamics is handled well on the Determinate Particle Speculation. The quantum state describes a suitable collection of systems. And the von Neumann dynamics is simply a matter of making an observation on one member of the collection, while the Schroedinger dynamics describes the evolution of the whole collection. On other speculations we can always resort to the heroic metaphysical hypothesis that there is a superposition of mental states corresponding to a superposition of brain states. In that case the von Neumann dynamics can be interpreted as describing an observation made by 'one of' the persons forming part of that superposition. However, there are also dualist accounts, such as that discussed in section III of this chapter.

Schroedinger's Cat poses no difficulty on the Determinate Particle Speculation. In some members of the collection the cat is intact, in others it is exploded. On the Wave Speculation we have the decidedly peculiar situation in which the feline wave is spread out, a superposition of intact and exploded cats. Intermediate is the mere failure of bivalence obtained on the Indeterminate Particle Speculation. It is neither true nor false that the cat is intact.

I do not consider a non-accidental correlation at a distance ('action-at-a-distance') to be a drawback, but I list the comparison because it is an interesting point, and because some readers may be unconvinced that action-at-a-distance is no problem. Likewise I do not consider any failure of the Converse of Weak Veridicality to be a drawback. Again I deny that the rejection of the Dyadic Programme is a defect, and I am happy about uninstantiated properties.

Unless we posit negative energies, the classical Principle of Energy Conservation is true only on the Wave Speculation. I leave it to the reader to decide if this is a significant advantage for the Wave Speculation.

On all three speculations we can ensure Weak Veridicality for suitable YNOs. We also have Statistical Veridicality or Mean Veridicality, as appropriate, for suitable observations. But we have to be careful in the case of momentum that we suitably describe what it is we are trying to measure.

The Wave Speculation scores well when it comes to superposition. However, on both it and the Indeterminate Particle Speculation we have a problem with mixtures, namely that different mixtures can be observationally indistinguishable (see the discussion in chapter 3).

The Schroedinger dynamics may be described as the straightforward determinate evolution of states on either the Wave or the Indeterminate Particle Speculations. But on the Determinate Particle Speculation, not only is it described by a messy equation, it involves a holistic statistical constraint on all systems. This is more complicated than we would like.

Both the Wave and the Indeterminate Particle Speculations require novel operations on properties, namely those needed to 'build up' polywaves, on the one hand, and indeterminate properties, on the other. And the chief conceptual innovations required by these two speculations are just the polywaves and the indeterminate properties. So we should not count both lines of the table as costs. I have listed the one cost twice to bring out what is involved. Moreover, while the Determinate Particle Speculation involves no novel operations, it does require a conceptual innovation, namely tied (and negative) propensities.

I have already mentioned that on the Determinate Particle Speculation the Schroedinger dynamics involves statistical constraints of a holistic kind, that is ones concerning *all* systems. A related difficulty is that we cannot even describe the quantum state in ways that are both intrinsic (i.e. making no reference to other systems) and occurrent (i.e. involving neither propensities nor dispositions). This is because the state is partly characterized by the propensities. A further difficulty with the Determinate Particle Speculation is that these propensities turn out to be relations not analysable into properties. While I welcome the prospect of unanalysable relations, I am alarmed by their high polyadicity. (For the system consisting of the whole universe, the polyadicity equals the total number of particles.)

I have noted that the Indeterminate History Variant requires negative masses. This variant seems superior to the other variants of the Indeterminate Particle Speculation because it enables us to give an occurrent description of the 'probability current', which plays an important role in the Indeterminate Particle Speculation. This raises the question of how reluctant we should be to posit negative mass (and negative energy). I leave this up to the reader, merely noting that we can give an account of spin in terms of rotation if we have negative mass.[1]

I sum up the various successes and difficulties at the bottom of the table of comparisons.

II FOUR RESPONSES TO THE COMPARISONS

What I have called the New Metaphysics requires freedom to speculate – and I have speculated. Should we draw any final conclusions? None of our speculations affected the argument, largely from Bell's Inequality against the Dyadic Programme. Hence it should definitely be rejected. Now let us consider the choice between the three speculations. One possibility here is that a little more thought or insight would have produced a far better speculation. Indeed I have a hunch that one is just round the corner. If so, maybe it will be partly inspired by my own speculations, in which case I shall be proud to have played my part. Or maybe I shall be humbled. In either case my comments on deciding between speculations will be out of place. So be it. But meanwhile I consider the speculations I have been able to provide.

I do not exclude either the agnostic response (i.e. not deciding between them), nor belief in one of the speculations and rejection of others. Both are rendered rationally permissible by the way every speculation is best in some respects. It is not as if there is a speculation which is best in all respects, or worst in all respects.

It might be objected, however, that we should reject *all* the speculations. This would not be because of a few difficulties, which merely amount to a greater complexity than we would like. Rather the case for rejecting all the speculations would, I think, be that there is too much of the ad hoc in all of them. Tied propensities, polywaves and indeterminate disjunction are all literally ad hoc hypotheses in the sense that they are posited only to handle some specified problem.

This response is too harsh. There is a place for the ad hoc provided it is kept to a minimum. Here we should distinguish between two judgements concerning belief and explanation. The first is that a hypothesis explains various data so successfully and with such economy that it is *presumptively* irrational not to accept it. This applies, I claim, to many scientific theories. The extent of my appeal to the ad hoc excludes that sort of judgement for the speculations we have considered. But it does not exclude a weaker judgement, namely:

> It is rationally permissible (but perhaps not rationally obligatory) to believe a speculation because of its explanatory power, provided one has fully considered available rival speculations.

I claim that the metaphysical speculations being considered are subject to *that* sort of judgement. Now, of course, a sceptic could deny that explanatory power is *any* guide for belief. But that *is* a sceptical position, and one I am not obliged to answer. Again, it could be objected that the explanations I have provided are just not good enough. That is not something that can be discussed. It is up to individual judgement. So I pass on. Yet again it might be objected that the literally ad hoc character of our speculations prevents them even being rationally permissible ones. Now that is a point which can be discussed. For I can produce examples of scientific speculations with a comparable resort to the ad hoc, and appeal to the reader to judge that it is rationally permissible to accept them. Two examples will suffice. After Leverrier and Adams had postulated the existence of Neptune, and before it was observed by Galle, it would have been rationally permissible to suspend judgement about its existence, but also rationally permissible to believe in its existence. Yet it was an ad hoc explanation in the sense that it was proposed to explain just a single phenomenon, the perturbation of the orbit of Uranus. Again, consider the theory of continental drift, and let us *pretend* that it was initially proposed only to explain the apparent fit between the continents. That would be literally ad hoc. Yet, in spite of the hostility with which the theory was initially received even with much more evidence, I judge that it would have been rationally permissible, but not rationally obligatory, to believe the theory as an explanation of a single phenomenon.

Provided, then, we are modest in the status we give to these speculations, the presence of the ad hoc, while undesirable, is not catastrophic. From this we may draw an important conclusion: the case against Realism has been greatly exaggerated. That is, there is no argument from quantum theory against the scientific realist tradition which I have entwined with the New Metaphysics. If there had been no tenable speculation about quantum theory which was realist about theoretical entities, then it could have been argued that it was irrational both to accept quantum theory and be a scientific realist. But I have provided three such speculations.

Another response is the pluralist one, which I now discuss. All three speculations have explanatory power. So why not accept two, or even all three? I begin by distinguishing Moderate Pluralism from one weaker and two stronger positions. The weaker position is merely the one implicit in what I have already said, namely that it is rationally permissible to accept any of the three speculations. The position I am examining, however, is that it is rationally permissible to accept more than one at the same time. One of the two stronger positions, which I am not advocating, is a relativist position. At his crudest, the relativist would say that each position is true for some people. That is by no means what I am considering, which is the position that all three speculations are just true *simpliciter*. Another position which I am not concerned with is a more radical Pluralism according to which 'anything goes'. I am operating with quite stringent constraints on theories. In particular I have rejected several speculations and variants of speculations on grounds which are not strictly conclusive.

I shall now briefly defend what I call the Pluralist Presumption,[2] namely, the presumption in favour of all the good explanatory theories. Then I shall discuss whether it is overcome.

First, let me qualify the Pluralist Presumption by stipulating that the theories being considered are suitably approximate and restricted. So instead of considering Newton's theory of gravitation we could consider the approximate inverse square law in a rather vaguely characterized set of circumstances. Having made this qualification, I now argue for the Pluralist Presumption in two ways. First let us consider the alternative. This is inference to *the* best explanation (provided, perhaps, that the best is significantly better than the next best). But what about all the marvellous theories no one has thought of? What degree of belief should we have that there is some better as yet undiscovered theory? To the extent that we do believe this, then we should be sceptical about inference to the best explanation. This problem is lessened, although not completely eliminated, if we hold the Pluralist Presumption. For in that case, the marvellous theory no one has thought of could be accepted without the best *available* one being rejected, unless they are incompatible. And the chance of an undiscovered theory being incompatible with the best available is less than the chance of one which is merely better.

The second argument for the Pluralist Presumption is to rely on a version of Smart's Cosmic Coincidence Argument (Smart, 1963, p. 39). We have various phenomena to explain. And we may suppose that a given theory, which is not *too* ad hoc, would explain these phenomena. Then, if the theory is not true it is a *cosmic coincidence* that the phenomena are *as if* it were true. Assuming we reject cosmic coincidences where we can, this establishes a presumption in favour of the truth of explanatory theories, regardless of whether they are best or tenth best.

A case can be made then for the Pluralist Presumption. Why has it not found more favour? Some of the hostility towards it is probably due to the confusion of it with the more radical, Feyerabendian, thesis that 'anything goes'. Much of the remaining hostility may be illustrated by considering as an example the supposition that both a wave and a particle speculation are correct. How could this be?

On the Wave Speculation the universe is one big polywave. A polywave is analysable, by means of the operations of superposition and composition, into monowaves. On the Particle Speculation the universe is a structured system of particles. Hence for the polywave to be the system of particles, the one nature, namely that of the whole physical universe, has to be analysable in two quite different ways. That, I take it, is the origin of much hostility to Pluralism. We tend to take for granted something which I call the Uniqueness of Analysis. Assuming, for simplicity,[3] that a complex nature can be analysed into simple properties and relations, then the Uniqueness of Analysis tells us that the set of properties and relations obtained in this way is unique. This claim has only to be made explicit for one to begin doubting it. To be sure, someone could claim just to know it a priori. But I make no such claim. And there can be no other grounds for holding it. I conclude that the one nature might well be analysed into simpler properties in several different ways.

The Pluralist Presumption is, however, just that, a *presumption*. And in the case of waves versus particles, this presumption is overcome. Here we must be careful not to confuse the wave *formalism*, discussed in chapter 3, with the Wave *Speculation*. Of course, the wave formalism is compatible with, say, the Indeterminate Particle Speculation. We just take the 'waves' to be disturbances defined using the 'probabilities' for location-cum-spin and the current. But on the Wave Speculation we give the wave a certain role to play – we ascribe to it causal powers. In particular, we associate a mass density to each disturbance to the field. So a wave has a density for mass spread out in space. By contrast, a system of determinate particles has its mass concentrated into a few points. And it is a mistake to think of even an indeterminate system of particles as having a spread-out mass. Rather it is indeterminate between various concentrated distributions of mass. Wave and Particle Speculations then, entail inconsistent accounts of the distribution of mass in the universe. So they are not compatible.

I shall argue, however, that the Determinate and Indeterminate Particle Speculations *are* compatible. We might suppose as much on the grounds that where the latter fails to specify the truth-value the former may be called on to provide it.

Recall that on the Determinate Particle Speculation we do not have to consider an *actual* collection: it suffices to consider a collection of *possible* systems. Now let us compare this with the Indeterminate History Speculation (a variant of the Indeterminate Particle Speculation). Each system in the collection of possible systems is specified by a history. So on either the Determinate or the Indeterminate Particle Speculation the quantum state at a given time corresponds to a suitable collection of world histories. The main difference between the two speculations is that on the Determinate Particle Speculation one of the determinate histories is the actual one, but on the Indeterminate Particle Speculation, the actual history is indeterminate between various determinate histories.

We can reconcile the two speculations by distinguishing *time-dependent* from *omnitemporal* actuality. And it is convenient here to consider the system consisting of the whole physical universe. An indeterminate history of the whole physical universe is the indeterminate disjunction of the members of some set of possible worlds, which for convenience I shall take to be a finite set. This set is then a large collection of possible systems of which just one, according to the Determinate Particle Speculation, is actual.

Now observation[4] results in a transition from one indeterminate world to another. The corresponding set of possible determinate worlds is smaller after the transition. We may say, then, that the state at time t is given by the set of all worlds which are *actual at time t*. Call this set A_t. Then, if s is an earlier time than t, $A_t \subseteq A_s$.

We may now characterize the omnitemporally actual worlds.[5] They are the worlds belonging to A_t for all t. They form the set $\bigcap A_t$. Call this A_∞. Nothing I have yet said guarantees that there is a unique member of A_∞. It could contain several members, or no members at all. If we assume that all the A_t are finite sets, or if we impose suitable topological constraints, then we may ensure that A_∞ is non-empty.[6] Further constraints would ensure that A_∞ has a unique member.[7]

We are now in a position to reconcile the Determinate and the Indeterminate Particle Speculations. The former treats as actual only the omnitemporally actual world, while the latter is concerned with time-dependent actuality.

There are some details which need to be mentioned. First, observation (and perhaps other effects of the mental) results in backwards causation *of a sort*. Suppose an observation is made at time t and that, as a result, A_t is a proper subset of A_s, where s is an earlier time. Now consider what happens at time s. If by that we mean what happens at time s in the (indeterminate) world actual at time s then it is not affected by the later observation at time t. But if we mean what happens at time s in the omnitemporally actual world, then that is affected by the later observation. It is affected because the observation is part of the process which picks out the omnitemporally actual world, and not because it in any way affects what goes on in one of the possible determinate worlds.

A second difference between the two speculations concerns the way the problem of negative probability was handled. On the Determinate Particle Speculation we attached no physical significance to the joint frequency distribution for position and (occurrent) momentum. Furthermore, in place of the dispositional momentum we considered a momentum analogue, the negative tied propensities. On the Indeterminate History Speculation this solution was not available. So we introduced negative mass.

In both cases we were guided by the need to make sense of a joint probability distribution which took negative values. Now we require only *one* way of making sense of this. Hence, if both the Determinate Probability and the Indeterminate History Speculations are correct, we do not require *both* solutions to this problem; either will do. What we have, then, are two variants of the Determinate-cum-Indeterminate Particle Speculation, based on the two different ways of handling negative probability.

Even if we rejected the Pluralist Presumption, we could consider what advantages and disadvantages the Determinate-cum-Indeterminate Speculation has, considered as a single hypothesis. First, it combines the explanatory power of the two speculations. For example, it handles the Two Dynamics Problem and Schroedinger's Cat as well as the Determinate Particle Speculation. But it does not have the problems of the Schroedinger dynamics which the Determinate Particle Speculation has. On the other hand, for the most part, it combines the ontological commitments of the two speculations. We have the novel operations on properties and new commitments to uninstantiated properties, all of which could have been avoided on the Determinate Particle Speculation. The exception to this is that we are not committed both to tied propensities and negative mass. For as I explained above, we can chose just one of these.

A more difficult question is whether the Determinate-cum-Indeterminate Particle Speculation is as materialist-friendly as the Determinate Particle Speculation. I think not. For on a materialist account it would be mysterious why there should be an increase of determinacy on observation. But without that the proposed reconciliation of interpretations cannot be brought about.

For what it is worth, I judge the Determinate-cum-Indeterminate Particle Speculation to be slightly superior to either speculation by itself. I am also inclined to favour it over the Wave Speculation, which I do not, however, entirely dismiss.

I have reached, then, the major conclusion of this work, namely that there are tenable realist speculations about quantum theory. Perhaps I should offer one final conjecture about the widespread resistance to such realist speculations. Here we should first note that none of the speculations I have given are as satisfactory as we might have hoped. As I have conceded, they are all tainted with the ad hoc. I suspect that there are two common tendencies which result in *excessive* reluctance to accept that which is even a little ad hoc. The first is a tendency to dismiss as impossible, or – worse – incoherent, that which is merely ad hoc or novel. Now I have no quarrel with the thinker who *on reflection* insists that he knows a priori that something (e.g. backward causation) cannot occur. What I object to is the unreflective dismissal of a speculation because of some minor blemish. The second, related, tendency is the vice of *perfectionism*. Because we cannot have a speculation quite as satisfactory as we would like (little gritty granules?), we tend to reject as worthless what is in fact a quite tenable speculation.

III PERSONS AND QUANTUM THEORY

One of the fundamental problems since the rise of modern science has been how persons fitted into the world as described by science. It is beyond the scope of this work to discuss the various proposals such as Cartesian Dualism, Idealism, Epiphenomenalism and Materialism, or to document the attempts of many philosophers, including Heidegger, Wittgenstein and Ryle, to work their way free from Cartesian presuppositions. My task is far more modest.

I am concerned with what difference quantum theory makes to this discussion.

At the outset, let me emphasize that I am not considering the impact which quantum theory has actually had. My interest is in the normative question of what impact it *should* have. Let us begin, then, by sketching the problems which classical mechanics poses for traditional accounts of persons. Here we are to suppose that there is a physical world made up of determinate particles each with position and momentum at all times; and these particles are governed by deterministic laws.

This classical-mechanical description of the world could be modified by including fields. But, for simplicity, let us stick with particles. That description poses no problem for Materialism. How could it? The physical world is described without any reference to anything characteristically personal. So, in one way or another, it is open to someone to speculate that persons are nothing more than parts of the physical world. The problems for Materialism are in no way due to classical mechanics.

On the other hand, classical mechanics would generate a couple of difficulties for those of us – dualists in the broad sense – who claim that persons are more than just parts of the physical world. To bring out these difficulties, let us call this something extra, whatever its nature, the *spiritual*. Perhaps as Descartes claimed, there is a spiritual substance. Or perhaps there are just spiritual properties. But in any case, the dualist holds that there are some spiritual items involved in being a person.

The first difficulty generated by classical mechanics I call the Threat of Epiphenomenalism, which I shall now describe. The action of the physical on the spiritual might be somewhat puzzling, but this puzzlement seems largely independent of our theory of the physical. However, the problem of how the spiritual affects the physical does depend on our theory of the physical. And it is especially acute in the context of classical mechanics. For if our bodies – and in particular our brains – operate in accordance with the laws of classical mechanics, there is an adequate explanation of our bodily movements in purely physical terms. Hence any affect of the spiritual would be quite redundant.

Let us make this more precise, by considering *causal overdetermination*. Sometimes one thing, *C*, causes another, *E*, even though *E* would have occurred anyway. I say that this is a case in which *E* has been *overdetermined*. For example, suppose the window is open and so is the door. A wind gets up. I see the door is about to slam shut, so I shut it gently. In that case, the door's being shut is overdetermined. I caused it, but it would have happened anyway. If classical mechanics were the correct description of the physical world, any effect of the spiritual on the material would seem to be one of causal overdetermination. The bodily movements would have happened anyway even without the effect of the spiritual. Now this is not quite Epiphenomenalism, which is the position that the spiritual has no effect on the material. But it is close enough to it to be a threat to the traditional belief that a person has a spiritual component which can affect the material. The case of overdetermination is incompatible with what we had in mind in stating the traditional position. With a slight abuse of terminology, I call this the Threat of Epiphenomenalism.

There are ways of meeting the Threat of Epiphenomenalism, even in the context of classical mechanics. The most obvious is to insist that the laws of nature, while deterministic in form are not in Armstrong's sense (1983, p. 147) *iron* laws. That is they can have 'exceptions'.[8] For then the action of the spiritual on the physical could well consist in 'exceptions' to the laws of classical mechanics.

There is another solution to the problem, different versions of which have been proposed by Vendler (1984, pp. 118–36) and myself (Forrest, 1985b). Consider various possible physical worlds, each obeying the deterministic laws. On this solution, which of them is the actual history depends on the spiritual. We may think of the history of the world as like railway tracks. Since the laws are deterministic there are no railway points, no branches in the tracks. If the spiritual resulted in an exception to these laws, it would be as if we picked up the locomotive and put it down on another track. But on the proposal now being considered, the spiritual affects which track the locomotive has always been on, and always will be on.

Neither of the above solutions is wholly satisfactory in the context of classical mechanics. Considerations of elegance and simplicity, which are admittedly never conclusive, favour laws which admit of no exceptions over ones which do. And the other solution is open to all the traditional objections to backward causation. In particular, we could not *affect* which history is ours if we have already *observed* which history is ours. And classical mechanics does not prevent that observation. So it is ad hoc to insist that there are obstacles preventing us observing which is our history. This recourse to the ad hoc would be a defect with the proposed solution.

Similar difficulties arise with the traditional belief that persons can freely choose to do one thing rather than another. If we reject the usual sorts of Compatabilism, such freedom apparently requires an exception to the deterministic laws. The problems with this are the same as those for the claim that the spiritual affects the physical. Indeed, since we would take the free choice itself to be in the realm of the spiritual, this is probably just a special case of the previous difficulty. So I shall treat the Threat of Determinism and the Threat of Epiphenomenalism as the same problem.

Let us turn to quantum theory. There might now be two threats to Materialism arising from quantum theory. The first threat is that the Determinate-cum-Indeterminate Particle Speculation might well be considered to have advantages over the Determinate Particle Speculation. But as I argued in section II, it is not available to the materialist. So perhaps the best available speculation about quantum theory is one which a materialist cannot hold. This would set up a tension between Scientific Realism and Materialism, which is the last thing materialists would have expected. I grant, though, this is very far from being a conclusive argument. In the first place, even if I am right that the Determinate-cum-Indeterminate Speculation is superior, considered by itself, materialists might reject it on the grounds that, so they say, other considerations favour Materialism. Secondly, materialists might reject the superiority of that speculation anyway.

The second possible threat to Materialism would occur if, for any reason, you reject the Determinate Particle Speculation. For on the other speculations, the materialist seems driven to the heroic metaphysics discussed in chapter 3. And that might be considered a disadvantage.

Provided, however, the Determinate Particle Speculation is tenable, as I believe it is, then the direct threat to Materialism is slight.

The significance of quantum theory for our understanding of persons is rather that it radically alters the problems faced by dualists. For provided either the Determinate or the Indeterminate Particle Speculation is satisfactory, the Threat of Epiphenomenalism and the Threat of Determinism can both be met without recourse to the ad hoc.

First, let us assume that the Determinate Particle Speculation is correct. That is the more interesting case because without it the materialist is going to be in trouble anyway. We have, then, various statistical constraints on the physical world which can be described by means of propensities. How does this ward off the threats of Epiphenomenalism and Determinism? We should resist the temptation to be too swift here. To say 'Classical Determinism is incorrect, so the spiritual is free to affect the material' is rather like saying 'God is dead, so we can do what we like'. Festina lente.

In this connection Smart (1963, p. 123) claims that the neurones in the brain are, though small, large enough to be treated classically to a good approximation. I shall grant this. So suppose the only difference resulting from quantum theory is that events which in classical mechanics were inevitable are now just extremely likely. This, apparently negligible, difference has in fact considerable consequences. For it rescues the first solution to the problem of the threat of Epiphenomenalism and Determinism, namely that in which it is proposed that the laws are not iron ones, but ones which admit of 'exceptions'. This is especially clear if we think of the laws as statistical constraints. This is one sort of probabilistic law (see Armstrong, 1983, pp. 126–36) and in fact it is the sort which is required on the Determinate Particle Speculation, as I have developed it. Such a law is not an iron law. Rather it is more like a nihil obstat law.[9]

Let me now clarify the distinction between *iron* and *nihil obstat* laws. A law that all *F*s are *G*s is, at least contingently, an iron law. For if the law had an exception, some *F* would not be a *G*. To say that the law that *F*s are *G*s is a nihil obstat law is therefore to deny that it is of the form: *All F*s are *G*s. Rather, it would be of the form: Anything which is an *F* is a *G* unless some additional factor prevents it being so. It is because the nihil obstat clause 'unless some additional factor prevents it being so' weakens the law, that considerations of theoretical elegance and simplicity favour iron laws.

How does this distinction apply to probabilistic situations? First, we could have an iron law that all *F*s are *G*s, where *G* is itself a propensity. However, I have argued that on the Determinate Particle Speculation, we should instead be considering statistical constraints. The typical form of one of these is that a high proportion of *F*s are *G*s. For the sake of definiteness, let us suppose it is that 99.99 per cent of *F*s are *G*s. Since the law no longer has either the form:

All *F*s are *G*s

or:

Any *F* is a *G* unless . . .

we cannot straightforwardly classify it either as an iron law or as a nihil obstat law. But the crucial point is that however high the proportion involved, a statistical constraint is more like a nihil obstat law (Any *F* is a *G* unless . . .) than an iron law. This can be brought out by considering the inference we make if we conclude that the spiritual cannot affect the material (except by way of overdetermination). We would be making an inference from there being a law connecting *F*s and *G*s, and there being an *F*, to there being a *G*. Here, *being an F* will be being in some specified physical state and *being a G* will be being in some specified, slightly later, physical state.

Compare these three inferences:

Inference 1:
All *F*s are *G*s.
This is an *F*.
So this is a *G*.

Inference 2:
Any *F* is a *G*, unless something prevents it being one.
This is an *F*.
So this is a *G*.

Inference 3:
99.99 per cent of *F*s are *G*s.
This is an *F*.
So this is a *G*.

The first inference is deductively valid. Hence if it were an iron law governing the physical world that all *F*s were *G*s the spiritual could intervene only if it could 'break the laws of logic'. By contrast, the second inference explicitly allows for intervention. If the laws governing the physical world were nihil obstat ones then, obviously, there is room for the intervention of the spiritual.

The crucial point is that the third inference, a proportional syllogism, is like the second rather than the first. The rationality of making a proportional syllogism depends on the context in which it is made. Consider the standard example of the urn with 99 white and 1 black ball. In many contexts it is rational to infer that some given ball – the one I shall pick out – is white. But suppose all the balls had just been in the sun. Then, from knowledge of elementary physics, you might well believe that a ball warmer than the others, which you can feel but not see, is black. Quite generally, the rationality of making a proportional syllogism is *defeasible*. And it is defeated if there is some further relevant factor.

It follows that a statistical law is like a nihil obstat law in allowing room for the spiritual to intervene in the physical world. So even if the probabilities governing the behaviour of the neurones are very near 1, it is still the case that something that would previously have been ad hoc, namely the possibility of 'exceptions' to laws, is now a consequence of its being the kind of law it is.

What, then, has happened to the disadvantage a theory has if it posits nihil obstat laws? That penalty has already been incurred by the Determinate Particle Speculation. Indeed, the kind of law it required was its chief disadvantage.

A second common objection to the claim that quantum theory 'rescues free will' (Brandt, 1959, pp. 511–12) is that by appealing to quantum theory all we do is replace determinism by randomism, and that if our acts are the result of random events in the brain they are no more free or characteristically personal than if they were determined by earlier events.

This objection would hold if we proposed a rather different sort of probabilistic law, namely a law that *F*s are *G*s where *G* is a propensity to ϕ, and where propensities are not analysed in terms of frequencies. Given a law of that sort, we might well think that to avoid randomism, we would require an 'exception' to the law that *F*s are *G*s. But I have argued that we can assume they are statistical constraints and so more like *nihil obstat* laws. And it is this *nihil obstat* character, not their probabilistic character, that is relevant. So I am not appealing to random events in the brain. Rather I am appealing to the fact that even what is likely to happen in the physical world is itself not completely determined by physical events, which leaves room for additional factors, such as the spiritual to determine it.

I prefer the account I gave of probabilistic laws because it avoids mysterious propensities. But suppose I granted that there is not much to choose between accounts of probabilistic laws. Even in that case, a dualist is quite free to rely on one of the accounts, precisely because it coheres better with Dualism.

Next, consider what happens on the Indeterminate Particle Speculation. This had several variants. I suggested that the most plausible was the Indeterminate History Speculation. But even if it were slightly less plausible in other respects, we could favour it if it fitted in better with our beliefs about persons.

Given the Indeterminate History Speculation we have an account of observation as increasing determinacy. It is reasonable to generalize this to any effect of the spiritual on the material, and it provides the dualist account promised in chapter 3. The spiritual affects the material by increasing the determinacy of the world. This is like the second of the two solutions which I stated in the context of classical mechanics. We think of the physical world as becoming more and more determinate as a result of the operation of the spiritual. But, unlike the solution proposed in the context of classical mechanics, we no longer need an ad hoc exclusion of the observation of which history will become actual. For quantum theory already excludes such observations. One manifestation of this is Heisenberg's Uncertainty Principle.

To sum up: On either the Determinate or Indeterminate Particle Speculations, quantum theory is more congenial to our traditional beliefs about persons than was classical mechanics. (On the Wave Speculation there is little difference.)

NOTES

1 See note 19, chapter 2.
2 See Forrest (1986, pp. 175–7) for an account of 'inference to the best explanation' which supports the Pluralist Presumption.
3 The alternative, suggested by Armstrong (1978, vol. II, p. 35), is that some properties are infinitely complex. In that case we can still make sense of rival analyses in the following way. We consider the sets of properties into which a given property can be analysed. Then the Uniqueness of Analysis entails that any two such sets are contained in a third such set.
4 And perhaps free actions (Forrest 1985b). But for the moment concentrate on observation.
5 Even so what is omnitemporally actually true is ambiguous. For to be true of all omnitemporally actual worlds might not be quite the same as

being true at some time of all worlds actual at that time. The first is to be true of all members of A_∞, the latter is to be true of all members of A_t for some *t*.

6 We require a topology on the set of all possible worlds such that the A_t are not merely closed but compact. This ensures that A_∞ is non-empty.

7 It would suffice to have a metric on the possible worlds such that the diameters of the A_t tend to zero as *t* increases. This establishes that there is at most one member of A_∞.

8 More precisely if we refer to the law as the law that *F*s are *G*s then there are exceptions to the generalization that all *F*s are *G*s.

9 Here my terminology is not Armstrong's (1983, p. 106), who calls them *oaken* laws.

References

Araki, Huzihiro and Yanase, Mutsuo M. (1960) 'Measurement of Quantum Mechanical Operators', *Physical Review*, 120, 622–6.

Aronson, J. L. (1971) 'On the Grammar of "Cause"', *Synthese*, 22, 414–30.

Armstrong, D. M. (1968) *A Materialist Theory of Mind*. London: Routledge and Kegan Paul.

—— (1973) *Belief, Truth and Knowledge*. Cambridge: Cambridge University Press.

—— (1978) *Universals and Scientific Realism*, vols I and II. Cambridge: Cambridge University Press.

—— (1983) *What is a Law of Nature?* Cambridge: Cambridge University Press.

Audi, Michael (1973) *The Interpretation of Quantum Mechanics*. Chicago: Chicago University Press.

Ayer, A. J. (1971) *Language, Truth and Logic*. Harmondsworth: Pelican.

Ballentine, L. E. (1970) 'The Statistical Interpretation of Quantum Mechanics', *Reviews of Modern Physics*, 42, 358–81.

Bell, J. S. (1964) 'On the Einstein–Podolsky–Rosen Paradox', *Physics*, 1, 195–200.

Beltrametti, Enrico and Cassinelli, Gianni (1981) *The Logic of Quantum Mechanics, Encyclopedia of Mathematics and its Applications, vol. 15* (ed. Gian-Carlo, Rota). Reading, Mass.: Addison-Wesley.

Bohm, D. (1951) *Quantum Theory*. Englewood Cliffs, NJ: Prentice Hall.

Bohm, D. and Bub, J. (1966) 'A Proposed Solution of the Measurement Problem in Quantum Mechanics by a Hidden Variable Theory', *Reviews of Modern Physics*, 38, 453–69.

Bohr, Niels (1949) 'Discussions with Einstein on Epistemological Problems in Atomic Physics'. In Paul Schlipp (ed.), *Albert Einstein: Philosopher-scientist*, Evaston, Ill.: The Library of Living Philosophers, pp. 199–242; reprinted in *Atomic Physics and Human Knowledge*, New York: John Wiley and Sons, 1958, pp. 32–66.

Born, Max (1949) *Natural Philosophy of Cause and Chance*. Oxford: Oxford University Press.

Born, Max and Wolf, Emil (1980) *Principles of Optics*, 6th edn. Oxford: Pergamon Press.

Boothby, William M. (1975) *An Introduction to Diffentiable Manifolds and Riemannian Geometry*. New York: Academic Press.

Brandt, Richard B. (1959) *Ethical Theory*. Englewood Cliffs, NJ: Prentice Hall.

de Broglie, Louis (1960) *Non-linear Wave Mechanics, A Causal Interpretation*. Amsterdam: Elsevier.

Cartwright, Nancy (1983) *How the Laws of Physics Lie*. Oxford: Oxford University Press.

Cohen, Leon (1966) 'Can Quantum Mechanics be Formulated as a Classical Probability Theory?', *Philosophy of Science*, 33, 317–22.

Corbett, J. V. and Hurst, C. A. (1978) 'Are Wave Functions Uniquely Determined by their Position and Momentum Distributions?', *Journal of the Australian Mathematical Society*, 20 (Series B), 182–201.

Crutchfield, J. P., Farmer, J. D., Packard, N. H. and Shaw, R. S. (1986) 'Chaos', *Scientific American*, 255 (no. 6), 38–49.

Das, Ashok and Melissinos, Adrian (1986) *Quantum Mechanics: A Modern Introduction*. New York: Gordon and Breach.

Davidson, Donald (1984) 'On the Very Idea of a Conceptual Scheme'. In *Truth and Interpretation*, Oxford, Oxford University Press, pp. 183–198; first published in *Proceedings and Addresses of the American Philosophical Association*, 47 (1974).

Davies, E. B. (1976) *Quantum Theory of Open Systems*. New York: Academic Press.

Derrida, Jacques (1982) *Margins of Philosophy*, trans. Alan Bass. Chicago: Chicago University Press.

Destouches, J. L. (1980) 'Using de Broglie's Physical Waves to Solve the Einstein–Podolsky–Rosen Paradox', *Nuovo Cimenta*, 58, B, 73–85.

Devitt, Michael (1984) *Realism and Truth*. Princeton, NJ: Princeton University Press.

Dummett, Michael (1978) *Truth and Other Enigmas*. London: Duckworth.

—— (1982) 'Realism', *Synthese*, 52, 55–112.

Eberhard, P. H. (1977) 'Bell's Theorem without Hidden Variables', *Nuovo Cimento*, 38, B, 75–9.

Einstein, A., Poldosky, B. and Rosen, N. (1935) 'Can Quantum-Mechanical Description of Physical Reality be Considered Complete?', *Physical Review*, 47, 777–80.

Everett, Hugh (1957) '"Relative State" Formulation of Quantum Mechanics', *Reviews of Modern Physics*, 29, 454–62.

Everett, Hugh et al. (1973) *The Many Worlds Interpretation of Quantum Mechanics: A Fundamental Exposition* (ed. B. S. De Witt and N. Graham). Princeton, NJ: Princeton University Press.

Fair, D. (1979) 'Causation as Flow of Energy', *Erkenntnis*, 14, 219–50.

Feigl, Herbert (1958) 'The "Mental" and the "Physical"'. In Herbert Feigl. Michael Scriven and Grover Maxwell (eds), *Studies in the Philosophy of Science*, 2, Minneapolis, University of Minnesota Press.

Feynman, R. P. (1948) 'Non-relativistic Quantum Mechanics', *Reviews of Modern Physics*, 20, 367–87.

—— (1949) 'The Theory of Positrons', *Physical Review*, 76, 167–98.

Fine, Arthur (1973a) 'The Two Problems of Quantum Measurement', In P. Suppes (ed.), *Proceedings of the Fourth International Congress for Logic, Methodology and Philosophy of Science*.

—— (1973b) 'Probability and the Interpretation of Quantum Mechanics', *British Journal of Philosophy of Science*, 24, 1–37.

Forrest, Peter (1985a) 'What Reasons do we have for Believing there are laws of Nature?', *Philosophical Inquiry*, 7, 1–12.

—— (1985b) 'Backward Causation in Defence of Free Will', *Mind*, 94, 210–17.

—— (1986) *The Dynamics of Belief: A Normative Logic*. Oxford: Basil Blackwell.

Forrest, Peter and Armstrong, D. M. (1987) 'The Nature of Number', *Philosophical Papers*, 16, 165–86.

van Fraassen, Bas, C. (1972) 'A Formal Approach to the Philosophy of Science'. In Robert G. Colodny (ed.), *Paradigms and Paradoxes*, Pittsburgh, University of Pittsburgh Press.

—— (1973) 'Semantic Analysis of Quantum Logic'. In Clifford A. Hooker (ed.), *Contemporary Research in the Foundations of Philosophy of Quantum Theory*, Dordrecht, Reidel, pp. 80–113.

—— (1980) *The Scientific Image*. Oxford: Oxford University Press.

—— (1982) 'The Charybdis of Realism: Epistemological Implications of Bell's Inequality', *Synthese*, 52, 25–38.

Ghirardi, G. C., Rimini, A. and Weber, T. (1980) 'A General Argument against Superluminal Transmission through the Quantum Mechanical Measurement Process', *Lettere al Nuovo Cimento*, 27, 293–8.

Gibbins, P. F. (1983), 'Quantum Logic and Ensembles'. In Richard Swinburne (ed.), *Space, Time and Causality*, Dordrecht, Reidel, pp. 191–205.

Gibbins, P. F. and Pearson, D. B. (1981) 'The Distributive Law in the Two-Slit Experiment', *Foundations of Physics*, 11, 792–803.

Goodman, Nelson (1954) *Fact, Fiction and Forecast*. London: Athlone Press.

Heidegger, Martin (1959) *An Introduction to Metaphysics*, trans. Ralph Mannheim. New Haven, Conn.: Yale University Press.

Heisenberg, Werner (1930) *The Physical Principles of the Quantum Theory*, trans. Carl Eckart and Frank C. Hugh. New York: Dover.

—— (1959) *Physics and Philosophy: The Resolution in Modern Science*. London: Allen and Unwin.

Hooker, Clifford A. (1971) 'Energy and the Interpretation of Quantum Mechanics', *Australasian Journal of Philosophy*, 49, 262–70.

—— (1972) 'The Nature of Quantum Mechanical Reality: Einstein versus Bohr'. In Robert G. Colodny (ed.), *Paradigms and Paradoxes: The Philosophical Challenge of the Quantum Domain*, Pittsburgh, University of Pittsburgh Press, pp. 67–302.

—— (1974), 'Systematic Realism', *Synthese*, 26, 409–97.

—— (1976), *The Logico-algebraic Approach to Quantum Mechanics*, vol. I. Dordrecht: Reidel.

—— (1978) (ed.), *Physical Theory as Logico-operational Structure*. Dordrecht: Reidel.

—— (1979), *The Logico-algebraic Approach to Quantum Mechanics*, vol. II. Dordrecht: Reidel.

Jarrett, Jon P. (1984) 'On the Physical Significance of the Locality Conditions in the Bell Arguments', *Noûs*, 569–90.

Jauch, J. M. (1968) *Foundations of Quantum Mechanics*. Cambridge, Mass.: Addison-Wesley.

Jauch, J. M. and Piron, C. (1963) 'Can Hidden Variables be Excluded in Quantum Mechanics?', *Helvetica Physica Acta*, 36, 827–37.

Kochen, Simon and Specker, E. P. (1967) 'The Problem of Hidden Variables in Quantum Mechanics', *Journal of Mathematics and Mechanics*, 17, 59–87.

Kuhn, T. S. (1970) *The Structure of Scientific Revolutions*. Chicago: Chicago University Press.

Landau, L. D. and Lifschitz, E. M. (1958) *Quantum Mechanics: Non-relativistic Theory*, trans. J. B. Sykes and J. S. Bell. London: Pergamon Press.

Landé, Alfred (1965) *New Foundations of Quantum Mechanics*. Cambridge: Cambridge University Press.

Lewis, David (1966), 'An Argument for the Identity Theory', *Journal of Philosophy*, 63, 17–25.

—— (1979), 'A Subjectivist's Guide to Objective Chance'. In Richard C. Jeffrey (ed.), *Studies in Inductive Logic and Probability*, vol. II, Berkeley, Ca, University of California Press.

Lewis, David (1983) 'New Work for a Theory of Universals', *Australasian Journal of Philosophy*, 61, 343–77.

Mackey, George W. (1963) *The Mathematical Foundations of Quantum Mechanics*. New York: W. A. Benjamin.

Mackie, J. L. (1973) *Truth, Probability and Paradox*. Oxford: Oxford University Press.

Maxwell, Grover (1962) 'The Ontological Status of Theoretical Entities', In Herbert Feigl and Grover Maxwell (eds), *Minnesota Studies in the Philosophy of Science*, vol. III, Minneapolis, University of Minnesota Press, pp. 3–77.

Mehra, Jagdish (1974) *The Quantum Principle: Its Interpretation and Epistemology*. Dordrecht: Reidel.

Mellor, D. H. (1971) *The Matter of Chance*. Cambridge: Cambridge University Press.

Messiah, Albert (1961) *Quantum Mechanics*, vol. I, trans. G. M. Temmer and J. Potter. Amsterdam: North Holland.

—— (1962) *Quantum Mechanics*, vol. II, trans. G. M. Temmer and J. Potter. Amsterdam: North Holland.

McGinn, Colin (1982) 'Realist Semantics and Content Ascription', *Synthese*, 52, 113–14.

von Neumann, John (1955) *Mathematical Foundations of Quantum Mechanics*, trans R. T. Beyer. Princeton, NJ: Princeton University Press.

Nozick, Robert (1981) *Philosophical Explanation*. Cambridge, Mass.: Harvard University Press.

Parsons, Terence (1980) *Nonexistent Objects*. New Haven, Conn.: Yale University Press.

Piron, C. (1976) *Foundations of Quantum Physics*. New York: W. A. Benjamin.

Place, U. T. (1956) 'Is Consciousness a Brain Process?', *British Journal of Psychology*, 47, 44–50.

Pool, James (1968a) 'Baer*-Semigroups and the Logic of Quantum Mechanics', *Communications in Mathematical Physics*, 9, 118–41.

—— (1968b) 'Semimodularity and the Logic of Quantum Mechanics', *Communications in Mathematical Physics*, 9, 212–28.

Popper, Karl R. (1982) *Quantum Theory and the Schism in Physics*. Totowa, NJ: Rowman and Littlefield.

Priest, Graham (1979a) 'The Logic of Paradox', *Journal of Philosophical Logic*, 8, 219–29.

—— (1979b) 'A Note on the Sorites Paradox', *Australasian Journal of Philosophy*, 57, 74–5.

Prior, Elizabeth (1985) *Dispositions*. Aberdeen: Aberdeen University Press.

Prosperi, G. M. (1971) 'Macroscopic Physics, Quantum Mechanics and Quantum Theory of Measurement'. In T. Bastin (ed.), *Quantum Theory and Beyond*, Cambridge, Cambridge University Press.

Putnam, H. (1965) 'A Philosopher Looks at Quantum Mechanics'. In Robert G. Colodny (ed.), *Beyond the Edge of Certainty*, Englewood Cliffs, NJ, Prentice Hall, pp. 75–101.

—— (1975) *Mathematics, Matter and Method, Philosophical Papers*, vol I. London: Cambridge University Press.

—— (1978) *Meaning and the Moral Sciences*. London: Routledge and Kegan Paul.

Redhead, Michael (1983) 'Nonlocality and Peaceful Coexistence', In Richard Swinburne (ed.), *Space, Time and Causality*, Dordrecht, Reidel.

—— (1987) *Incompleteness, Nonlocality, and Realism: A Prolegomenon to the Philosophy of Quantum Metaphysics*. Oxford: Clarendon Press.

Rorty, Richard (1979) *Philosophy and the Mirror of Nature*. Princeton, NJ: Princeton University Press.

Routley, Richard (1979) *Exploring Meinong's Jungle and Beyond*. Canberra: Departmental Monograph, Philosophy Department, Research School of Social Sciences, Australian National University.

Schroedinger, Erwin (1952) 'Are There Quantum Jumps? Parts I and II', *British Journal for the Philosophy of Science*, 3, 109–23, 233–42.

Smart, J. J. C. (1959) 'Sensations and Brain Processes', *Philosophical Review*, 68, 141–56.

—— (1963) *Philosophy and Scientific Realism*. London: Routledge and Kegan Paul.

Suppes, Patrick (ed.) (1973) *Logic, Methodology and Philosophy of Science: Proceedings of the 4th International Conference, Bucharest 1971*. New York: American Elsevier.

Vendler, Zeno (1984) *The Matter of Mind*. Oxford: Oxford University Press.

Wigner, Eugene P. (1932) 'On the Quantum Correction for Thermodynamic Equilibrium', *Physical Review*, 40, 749–59.

—— (1952) 'Die Messung Quantenmechanischer Operatoren', *Zeitschrift für Physik*, 133, 101–8.

—— (1962) 'Remarks on the Mind-body Question'. In I. J. Good (ed.), *The Scientist Speculates: An Anthology of Partly Baked Ideas*, London, Heinemann.

—— (1963) 'The Problem of Measurement', *American Journal of Physics*, 31, 6–15.

—— (1971), 'Quantum-mechanical Distribution Functions Revisited'. In Wolfgang Yourgrau and Alwyn van de Merwe (eds), *Perspectives in Quantum Theory: Essays in Honor of Alfred Landé*, Cambridge, Mass., MIT Press.

Wyszecki, G. and Stiles, W. S. (1967) *Color Science: Concepts and Methods, Quantitative Data and Formulas*. New York: John Wiley and Sons.

Glossary

References given in brackets are to chapter and section number within the text.

action-at-a-distance (3.7)
Strictly, a direct (q.v) causal connection for which, however, cause and effect are spatially separated; more loosely, any direct (q.v) non-accidental correlation between spatially separated items.

actual, omnitemporally (8.2)
That which is actual relative to all times.

actual, time-dependent (8.2)
That which is actual relative to some given time.

Argument from Superposition (6.4)
The argument in favour of the Wave Speculation over the Fluid Variant (q.v) and over the Hyperspace Variant (q.v), based on the desideratum that we explain superposition (q.v).

associated location
See *location, associated.*

associated momentum
See *momentum, associated.*

backward causation (3.1)
Backward causation occurs if the cause is later than the effect.

bivalence (1.3)
Bivalence holds for a certain class of statements if there are none which are both true and false and none which are neither true nor false.

bosons (2.4)
Particles (q.v) whose spin (q.v) is a non-negative integer in units in which Planck's constant equals 2π. See also *fermions.*

collapse of the wave packet (2.1)
The von Neumann dynamics (q.v).

commuting YNOs
See *observation, commuting Yes/No.*

complementarity (4.3)
Bohr's claim that in quantum theory we need *both* of two incompatible models or pictures (e.g. waves and particles, position and momentum).

Complexity Problem (2.1)
The problem of the excessive theoretical complexity which seems to occur because we have both Schroedinger dynamics (q.v) and von Neumann dynamics (q.v).

Composite Systems Dilemma (3.8)
Sometimes the components (e.g. the electron and proton) are not in a pure state (q.v) even though the composite system (e.g. a hydrogen atom) is. The dilemma is whether to say the components are not in a quantum state or to say they are in a mixed state (q.v).

Conservation of Energy
See *Energy Conservation.*

Conservative Argument against waves (6.4)
The argument in favour of the Fluid Variant (q.v) on the basis that it is a more conservative speculation than the Wave Speculation.

Conservative Presumption (4.2)
The presumption in favour of traditional concepts and beliefs. (See also *Radical Privilege.*)

Converse of Weak Veridicality (3.3)
The thesis that whenever a Yes result for a given YNO (q.v) would be true of the system even prior to observation, then the system is in an eigenstate (q.v) for that YNO.

current (5.6, 6.1, 7.1)
A vector quantity associated with each location of a particle (q.v), definable in terms of the joint probabilities (q.v) for position and momentum as the mean momentum for that position. It may be characterized even without a joint distribution, as the momentum the particle would have if it were to be observed at that location.

decomposable state
See *state, decomposable.*

Demarcation Problem (3.2)
The problem of when Schroedinger dynamics (q.v) applies and when von Neumann dynamics (q.v) applies.

determinate property
See *property, determinate.*

determinate value (of position or momentum) (1.3)
A sharp or precise value, that is one given accurately by a real number.

Determinism, Threat of (8.3)
The problem that the laws governing the physical world seem to leave no scope for free acts. See also *Epiphenomenalism, Threat of.*

diachronic unity
See *unity, diachronic.*

direct (3.7)
Given any class of relations, *C*, item *a* and item *b* are said to be

directly related by a member of the class, if there is no item x such that a and b are related by a member of C *because* a and x are so related and x and b are so related. For example, a directly causes b (i.e. is directly related by a member of the class of causal relations) if a causes b and there is no x such that a causes b *because* a causes x and x causes b.

disjunction, indeterminate (7.2)

An operation analogous to, but not the same as, ordinary disjunction, which forms finitely indeterminate properties (q.v) out of determinate properties. For example, the property of having a given finitely indeterminate history (q.v) is the indeterminate disjunction of the properties of having various determinate histories.

dispositional momentum

See *momentum, dispositional.*

Dyadic Programme (3.7)

The programme of describing the physical world without recourse to any relations other than dyadic ones.

dynamics

See *von Neumann dynamics*; *Schroedinger dynamics.*

eigenstate (2.1)

Elliptical for 1-eigenstate. A system is in an eigenstate for a YNO (q.v) if the von Neumann dynamics (q.v) gives a Yes answer to the YNO with probability one.

Energy Conservation, Classical Principle of (3.7)

The principle that the total energy of an isolated system is conserved.

Energy Conservation, quantum analogue (3.7)

For any range of values the energy could take, consider the YNO (q.v) of whether the energy is in that range. And suppose that if that YNO were performed at time t on an isolated system then the probability of a Yes result would be P_t. Then if the YNO is not performed, but instead the system evolves according to the Schroedinger dynamics (q.v), the principal states that P_t will not change with time.

Epiphenomenalism, Threat of (8.3)

The problem that the laws governing the physical world would seem to leave no room for the spiritual to affect the physical, except in so far as the physical is overdetermined by the physical and the spiritual. See also *Determinism, Threat of.*

fermions (2.4)

Particles (q.v) whose spin (q.v) is a non-negative integer plus one-half in units in which Planck's constant equals 2π. See also *bosons.*

field, monadic (6.1)

An assignment to each point in space of a (structured) set of properties, called the field at that point. (In simple cases, this set of properties is the same for all points.) At most one property from the set is instantiated at the point in question. This is called the *value* of the field at the point. By convention, if no property is instantiated we say the field has value zero. See also *wave.*

field, polyadic (6.2)

An n-adic field is the assignment (to each ordered n-tuple of points) of a (structured) set of n-adic relations, called the field for that n-tuple. At most one relation is instantiated by the points in the n-tuple in the order in which they occur. It is called the *value* of the field for the n-tuple of points. See also *polywave.*

field, strength or intensity of (6.1)

A quantity measured by non-negative real numbers which is determined by the value of a field at a given point. Typically we would take the square of the field strength to have the units of energy.

field-value (6.1)

A synonym for the value of the field. See *field, monadic*; *field, polyadic.*

finitely indeterminate property

See *property, indeterminate.*

Fluid Variant (6.4)

A variant of the Wave Speculation in which a 'particle' is likened to a fluid.

Heisenberg's Uncertainty Principle (2.1)

The principle that the product of the variances (q.v) for the observations of position and of momentum must be at least 1/4 in units in which Planck's constant equals 2π.

hidden state

See *state, hidden.*

hidden variable (5.1)

An extra parameter introduced to specify various hidden states (q.v).

history (7.4)

A determinate history of a particle is the assignment to each time of a determinate location (and spin) for that particle. A finitely indeterminate history is the finite indeterminate disjunction (q.v) of determinate histories. An infinitely indeterminate history is a suitable limit (q.v) of finitely indeterminate histories.

Hyperspace Variant (6.4)

A variant of the Wave Speculation which simplifies the set of field-values (q.v) by positing extra dimensions for space.

Ignorance Interpretation of Mixture (2.5)

The interpretation of mixed states (q.v) according to which a simple system can be in a mixed state because the mixed state represents the various subjective probabilities (q.v) that the system is in various pure states (q.v).

π-implication (7.2)

A property (or relation) π-implies a predicate if, of necessity, the predicate applies to whatever item has the property (or whatever items are related by the relation).

π-inconsistency (7.2)
A property (or relation) is π-inconsistent with a predicate if, of necessity, the predicate applies to no item with the property (or no items related by the relation.)

indeterminate disjunction
See *disjunction, indeterminate.*

indeterminate property
See *property, indeterminate.*

Individual Incoherence, Threat of (3.2)
The problem that the Schroedinger dynamics (q.v) and the von Neumann dynamics (q.v) seem to put incompatible constraints on a system being measured by means of a piece of apparatus. See also *Statistical Incoherence.*

infinitely indeterminate property
see *property, indeterminate.*

intrinsic angular momentum (2.3)
Usually known as spin. A component of the angular momentum a particle has about a point, which does not depend on the point considered. It is not usually attributed to any literal rotation of the particle about an axis. See also *spin-up and spin-down.*

iron law
See *law, iron.*

Irreducibility of Composite Systems (3.6)
The thesis that a composite system in a pure state, not decomposable into pure states (q.v) of the components, is not decomposable into *any* quantum states (q.v) of its components. See also *Composite Systems Dilemma.*

joint probabilities (2.6)
To say that there are joint probabilities for position and momentum is to say that there is a physically significant assignment of probabilities not just to sets of position and to sets of momenta but to sets of position/momentum pairs.

Kantian speculation (3.3)
The speculation that our knowledge of the physical world is restricted to the relations (chiefly causal) it has with us, the observers. See also *semi-Kantian speculation.*

law, iron (8.3)
A law admitting of no exceptions. See also *law, nihil obstat.*

Law, nihil obstat (8.3)
A law of the form: Any *F* is a *G* unless . . ., where some exceptional circumstances such as outside interference are specified.

limits of finitely indeterminate properties (7.2)
To each finitely indeterminate property (q.v), there is associated a

'probability' distribution. We may posit a limit operation on the indeterminate properties in such a way that there is a 'probability' distribution associated with the limit of finitely indeterminate properties, and that distribution is itself the limit of the associated probability distributions.

locality requirement

See *strong locality*.

location (1.5)

Synonym of *position*.

location, associated (6.1, 7.1)

On the Wave Speculation the associated location is the position of the centre of mass of the wave. On the Indeterminate Particle Speculation it is the 'mean' for the 'probability' distribution for location. On either speculation it equals the mean value for the results of observing the location.

Many Worlds Theory (3.2)

The speculation, based on the work of Everett, that there is a large collection of worlds, not just the one physical world we usually think of.

Mean Veridicality

See *Veridicality, Mean*.

Methodological Atomism (3.7)

The assumption that the physical world can be understood by analysing composite systems into their components and the relations between them.

momentum, associated

A determinate vector associated with a particle even if it has no determinate momentum. It is the mass times the rate of change of the associated location (q.v).

mixed state

See *state, mixed*.

mixture (2.5)

Consider two large collections of systems which are in quantum states (q.v) s_1 and s_2. If copies of these collections are (literally) mixed in proportions λ to $(1-\lambda)$, then the resulting collection is in a state which is the mixture of s_1 and s_2 in proportions λ to $(1-\lambda)$. A single system might also be said to be in a quantum state which is a mixture of s_1 and s_2, if a large collection of systems in that state is, according to quantum theory, observationally indistinguishable from that described above. See also *state, mixed; state, pure*.

momentum, dispositional (1.5)

A property of a particle describable as a disposition or tendency to change location. In classical mechanics this disposition will be manifested (unless there are infinite forces) by a rate of change of location v. The magnitude of the momentum is then mv, where m is the mass of the particle.

momentum, occurrent (1.5)

The mass times the rate of change of location.

negation of a Yes/No observation (2.1)
Given any YNO (Yes/No observation – q.v) A, there is another YNO, $\bar{A}$, which gives the result Yes just in case A gives No and vice versa. This is said to be a *negation* of A.

von Neuman dynamics (2.1)
The dynamics which constrains a system when it is observed using a YNO (q,v). See also *Schroedinger dynamics.*

nihil obstat law
See *law, nihil obstat.*

Non-uniqueness Problem (3.6)
The problem that two different mixtures (q.v) of pure states (q.v) can result in the same mixed state (q.v).

observable (2.1)
A quantity which can be measured to any given accuracy by means of enough commuting YNOs (q.v).

observation, commuting Yes/No (2.6)
Two or more YNOs (Yes/No observations) (q.v) are said to *commute* (or to be simultaneously observable) if the probability of a Yes result to any of them is independent of which, if any, of the others have been made.

observation, Yes/No (2.1)
A Yes/No observation (YNO) is an observation of whether or not something is the case, and the making of which satisfies the von Neumann dynamics (q.v). The only possible results are Yes or No.

orthogonal states
See *states, orthogonal.*

particle (5.1)
Something with either a point location or of small diameter, which cannot have more than one location, and whose location can change with time.

Phase Angle Argument (6.4)
The argument against the Wave Speculation based on the way the wave function underdetermines the quantum state (q.v) by a phase angle.

Pluralist Presumption (8.2)
The presumption that all hypotheses which provide satisfactory explanations are correct. It is overcome if not all the hypotheses are compatible.

polywave (6.2)
An n-adic polywave is a smooth assignment of some value in the n-adic field (q.v) to each n-tuple of points. See also *wave.*

privileged class (4.4)
The class of distributions of frequencies for which a *tied propensity* (q.v) is directly (q.v) manifested.

probability density (2.1)
A function (which is unique if we may assume it is continuous) assigning

a real number to each value of some quantity (e.g. position or momentum) of interest, such that for any suitable (i.e. measurable) set X of those quantities, the integral of the probability density over X equals the probability assigned to X (e.g. the integral of the probability density for position over a region V is the probability assigned to that region). See also *probability distribution*.

probability distribution (7.1)

A probability distribution (or measure) is a coherent assignment of probabilities to suitable (i.e. measurable) sets of values of some quantity of interest (e.g. position or momentum). See also *probability density*; *joint probability*.

probability measure

See *probability distribution*.

probability, subjective (5.2)

A numerical measure of the degree of strength of belief.

Problem of Two Dynamics

See *Two Dynamics, Problem of.*

Projection Postulate, Strong (2.5)

The thesis that (suitably idealized) observations are governed by von Neumann dynamics (q.v).

Projection Postulate, Weak (2.5)

The thesis that the distribution of quantum states in a large collection after (a suitably idealized) observation is that which would result from the von Neumann dynamics (q.v).

propensity (4.4)

A propensity to ϕ of degree α is a property that can be described dispositionally by saying that typically in a large collection of items with this propensity (and not subject to outside interference) a proportion (approximately) α will ϕ.

propensity, relative (4.4)

A propensity (to degree α) to ϕ in circumstance C is a property which an item can have even if not in circumstance C, but which in circumstance C manifests itself as a propensity (to degree α) to ϕ. See also *propensity, tied*.

propensity, tied (4.4)

A property that can be described dispositionally by saying that it will directly (q.v) manifest itself like a *relative propensity* (q.v) but only for a collection of items with the appropriate distribution for the various *circumstances*. For example, a tied propensity for change of location is only directly manifested by a large collection of particles with the appropriate distribution for location. (In this example, the *circumstance* of a relative propensity to charge location to q_2 is the initial location q_1.)

propensity-pure collection (5.6)
A large collection of particles each with the same (tied) propensities for change of location. See also *standard propensity-pure collection.*

property, determinate (7.2)
A property analysable neither as the finite indeterminate disjunction (q.v) of other properties nor as the limit (q.v) of such.

property, indeterminate (7.2)
A *finitely* indeterminate property is one analysable as the finite indeterminate disjunction (q.v) of determinate properties. An *infinitely* indeterminate property is a property which is neither determinate nor finitely indeterminate. So it is analysable as the limit (q.v) of finitely indeterminate properties.

pure state
See *state, pure.*

Radical Privilege (4.2)
The right to ignore the Conservative Presumption (q.v) in those cases where what is traditionally believed cannot be observed.

Reducibility of Composite Systems (3.8)
Negation of *Irreducibility of Composite Systems* (q.v).

Redundancy/Interaction Dilemma (5.5)
A dilemma, one horn of which is that the frequencies for location are redundant in the dynamics of a collection of systems, the other that the members of the collection seem to interact.

Schroedinger dynamics (2.1)
The dynamics constraining a quantum system which is neither being observed nor subject to any other outside interference. See also *von Neumann dynamics.*

semi-Kantian speculation (4.5)
The speculation that what we know of the physical world is restricted to the propensities (q.v) of various items to produce various results when observed. See also *Kantian speculation.*

simultaneous observation (2.6)
Two or more quantities observable by means of commuting YNOs (q.v) are said to be simultaneously observable.

spin
See *intrinsic angular momentum.*

spin-up and spin-down (2.3)
Consider a particle of spin one-half (in units in which Planck's constant equals 2π). For any direction k in space, there is a YNO of whether the spin is *up* with respect to k. The eigenstates (q.v) for that YNO are called spin-up states (with respect to k). The eigenstates for the negation of that YNO are called spin-down states (with respect to k).

standard propensity-pure collection (5.6)
A propensity-pure collection (q.v) in which the distribution for frequencies belongs to the privileged class (q.v) for the tied propensities (q.v).

state (1.7)
A member of a 'space' of states (q.v). The 'space' of states being considered should be clear from the context. See also: *state, decomposable*; *state, hidden; state, mixed; state, pure; state, quantum; states, orthogonal; sub-states.*

state, decomposable (2.4)
A quantum state (q.v) for a system of several particles for which each particle is in a pure state for a single particle system.

state, hidden (5.1)
A sub-state (q.v) of a quantum state (q.v).

state, mixed (2.5)
A quantum state (q.v) which can be expressed as a mixture (q.v) of other quantum states. See also *state, pure.*

state, pure (2.5)
A quantum state which cannot be expressed as a mixture (q.v) of other quantum states (q.v). See also *state, mixed.*

state, quantum (1.7)
Those states either of single systems or of large collections of systems which are considered in quantum theory. See also *states, space of.*

states, orthogonal (2.1)
Two states are said to be orthogonal if the 'angle' between them is 90°, or equivalently if they are represented by orthogonal vectors in the Hilbert space. In the three-dimensional analogue, orthogonal states are represented by rays making 90° with each other.

states, 'space' of (2.1)
A 'space' of states for a system (or collection of systems) is a set of properties (typically with a structure, i.e. with relations between the properties), such that the system (or collection of systems) must have one of the properties (and so 'be in one of the states') but cannot have more than one. There is no requirement that a given system (or collection of systems) have a unique 'space' of states. But unless otherwise indicated, I consider the 'space' of quantum states (q.v).

Statistical Incoherence, Threat of (3.2)
The problem that the Schroedinger dynamics (q.v) and von Neumann dynamics (q.v) seem to put incompatible constraints on the frequencies of various changes in a large collection being measured using copies of a given piece of apparatus. See also *Individual Incoherence, Threat of.*

Statistical Interpretation of Mixture (2.5)
The thesis that all mixed states (q.v) could be literally obtained as mixtures, that is that they are states of large collections not of single systems, and are describable by means of a frequency distribution for the pure states

(q.v) of the systems themselves. See also *Ignorance Interpretation of Mixture; Sui Generis Interpretation.*

Statistical Veridicality
See *Veridicality, Statistical.*

strength of field
See *field, strength or intensity of.*

Strong Locality (3.7)
The requirement that the results of making two spatially separated YNOs on a system are probabilistically independent relative to the state of the system prior to the observations.

Strong Veridicality
See *Veridicality, Strong.*

Sub-state (1.7)
Consider a system (or collection of systems) with two 'spaces' of states. A state *u* in one 'space' is said to be a sub-state of a state *v* in the other 'space' if, of physical necessity, every system in *u* is also in *v*.

Sui Generis Interpretation of Mixture (2.5)
The thesis that the operation of mixture (q.v) can be applied to single systems but, as applied to them, is unanalysable.

Superposition (2.1)
An operation on quantum states (q.v) which – unlike mixture (q.v) – produces pure states (q.v) out of pure states. A superposition of two states can be thought of as 'between' those states. Typically, we consider a superposition of two orthogonal states (q.v). In that case, we may think of the orthogonal states as like antipodal points on a sphere. The various non-trivial superpositions then correspond to all the other points on the sphere. See also *Argument from Superposition.*

Superselection Rule (2.1)
In the atypical situation in which there are no non-trivial superpositions (q.v) of two quantum states (q.v) (i.e. no superpositions other than the two states themselves), we say there is a superselection rule.

Synchronic Unity
See *Unity, Synchronic.*

Threat of Determinism
See *Determinism, Threat of.*

Threat of Epiphenomenalism
See *Epiphenomenalism, Threat of.*

Threat of Individual Incoherence
See *Individual Incoherence, Threat of.*

Threat of Statistical Incoherence
See *Statistical Incoherence, Threat of.*

Tied Propensity
See *Propensity, Tied.*

Two Dynamics, Problem of (3.2)
The problem that physical systems are governed both by the Schroedinger dynamics (q.v) and the von Neumann dynamics (q.v). See also: *Complexity Problem; Demarcation Problem; Individual Incoherence, Threat of*; *Statistical Incoherence, Threat of.*

Uncertainty Principle
See *Heisenberg's Uncertainty Principle.*

Uniformity Thesis (2.5)
The thesis that the interpretation of mixture (q.v) should depend on the account given of pure states (q.v). See also *Ignorance Interpretation*; *Statistical Interpretation*; *Sui Generis Interpretation.*

Unity, Diachronic (6.5)
Particles (q.v) are said to have diachronic unity if when two or more particles collide, resulting in the same number of particles and of the same kinds, then every one of the resulting particles is *the same as* one of the earlier particles.

Unity, Synchronic (6.5)
Particles (q.v) are said to have synchronic unity if there is a difference between a particle with a given history (q.v) and having several properties (e.g. mass, change, spin) on the one hand, and several particles all with that history and each with just one of the properties, on the other.

variance (2.1)
If there is a probability distribution for some quantity X, then we may consider its mean X_m and the square of the variation from the mean, $(X-X_m)^2$. The mean of $(X-X_m)^2$ is called the *variance.*

veridicality (3.3)
An observation (q.v) is said to be veridical if the result of observation equals the value of the observable prior to observation. See also *Converse of Weak Veridicality*; *Veridicality, Mean*; *Veridicality, Statistical*; *Veridicality, Strong*; *Veridicality, Weak.*

Veridicality, Mean (3.3)
Mean Veridicality holds for a type of observation if the mean value derived from the probability distribution for the results of observation equals the value of that which is being measured prior to observation.

Veridicality, Statistical (3.3)
Statistical Veridicality holds for a type of observation if the quantum state specifies a probability (or frequency) distribution for the values of the observable prior to observation which is the same (or approximates) the distribution for the results of the observation. See also *Veridicality, Universal Statistical.*

Veridicality, Strong (3.3)
A type of observation is said to be *strongly veridical* if all observations of that type are veridical (q.v).

Veridicality, Universal Statistical (5.1)
Universal Statistical Veridicality is said to hold if all types of observation considered in quantum theory are statistically veridical (q.v).

Veridicality, Weak (3.3)
A YNO (q.v) is said to be *weakly veridical* if it is veridical (q.v) whenever the state of the system is an eigenstate (q.v) for the YNO.

wave (6.1)
A smooth assignment to points in space of values in some monadic field (q.v). See also *polywave*.

YNO (Yes/No observation)
See *observation, Yes/No*.

Index